LANDSCAPES OF A NEW CULTURAL ECONOMY OF SPACE

Landscape Series

Volume 5

Series Editors:

Henri Décamps,
Centre National de la Recherche Scientifique,
Toulouse, France

Bärbel Tress,
Aberdeen University,
Aberdeen, United Kingdom

Gunther Tress,
Aberdeen University,
Aberdeen, United Kingdom

Aims & Scope:

Springer's innovative Landscape Series is committed to publishing high quality manuscripts that approach the concept of landscape from a broad range of perspectives. Encouraging contributions on theory development, as well as more applied studies, the series attracts outstanding research from the natural and social sciences, and from the humanities and the arts. It also provides a leading forum for publications from interdisciplinary and transdisciplinary teams.

Drawing on, and synthesising, this multidisciplinary approach the Springer Landscape Series aims to add new and innovative insights into the multidimensional nature of landscapes. Landscapes provide homes and livelihoods to diverse peoples; they house historic – and prehistoric – artefacts; and they comprise complex physical, chemical and biological systems. They are also shaped and governed by human societies who base their existence on the use of the natural resources; people enjoy the aesthetic qualities and recreational facilities of landscapes, and people design new landscapes.

As interested in identifying best practice as it is in progressing landscape theory, the Landscape Series particularly welcomes problem-solving approaches and contributions to landscape management and planning. The ultimate goal is to facilitate both the application of landscape research to practice, and the feed back from practice into research.

LANDSCAPES OF A NEW CULTURAL ECONOMY OF SPACE

Edited by

Theano S. Terkenli

University of the Aegean,
Lesbos, Greece

and

Anne-Marie d'Hauteserre

University of Waikato,
Hamilton, New Zealand

 Springer

A C.I.P. Catalogue record for this book is available from the Library of Congress.

ISBN-10 1-4020-4095-4 (HB)
ISBN-13 978-1-4020-4095-5 (HB)
ISBN-10 1-4020-4096-2 (e-book)
ISBN-13 978-1-4020-4096-2 (e-book)

Published by Springer,
P.O. Box 17, 3300 AA Dordrecht, The Netherlands.

www.springer.com

Cover photograph by Bärbel Tress and Gunther Tress

Printed on acid-free paper

Foreword by the series editors

The underlying motivation behind the Springer Landscape Series is to provide a much-needed forum for dealing with the complexity and range of landscape types that occur, and are studied, globally. At the same time it is crucial that the series highlights the richness of this diversity – both in the landscapes themselves and in the approaches used in their study. Moreover, while the multiplicity of relevant academic disciplines and approaches is characteristic of landscape research, we also aim to provide a place where the synthesis and integration of different knowledge cultures is common practice.

Landscapes of a New Cultural Economy of Space is the fifth volume of the series. Focusing on the transformations and changes in human life that influence landscape development, the book presents 'landscape' as the interface of human-environment interrelationships where the different processes of change are perceived and expressed. Theano Terkenli and Anne-Marie d'Hauteserre use this approach as a way to introduce a collection of contemporary discussions in landscape research and geography which look specifically at the cultural transformation and representation of landscapes. The subsequent chapters then present those processes determining cultural economies of space in different parts of the globe under the concepts of 'enworldment', 'unworldment', 'deworldment' and 'transworldment'.

In each case the authors offer inspiring and discursive reflections on an emerging, socially defined pattern of landscape. We recommend the book to students and researchers dealing with contemporary challenges in the economic and cultural representation of landscape, as rooted in social and human geography and other landscape-related disciplines.

Toulouse and Aberdeen, November 2005

Henri Décamps
Bärbel Tress
Gunther Tress

Contents

ACKNOWLEDGMENTS

The editors of this collective volume are especially grateful to Springer and formerly Kluwer Academic Publishers, as well as their editorial teams in Dordrecht, for their constant support, patience, encouragement and courtesy in bringing this effort to completion. In particular, Helen Buitenkamp, Sandra Oomkes and Ria Kanters, who coordinated the production of the book. Our thanks are equally addressed to the Landscape Series Editors for their committed and enthusiastic guidance of the project in its final stages: Bärbel Tress, Gunther Tress and Henri Décamps. We also wish to thank Theano's technical assistant, Olga Mironopoulou, for her unfailing, efficient and professional undertaking of the technical part of the project. Institutional gratitude goes to our departments, Department of Geography, University of the Aegean, Lesvos and Department of Geography, Tourism and Environmental Planning, University of Waikato, for providing context and support to our academic efforts so far.

The editors of this volume are mutually appreciative of working together on this project, in what has been a thoroughly inspiring, spirited and impeccable cooperation. Last, but not least, our ultimate thanks belong to the chapters' authors for their outstanding contributions to this work and for readily and creatively adopting our vision of the subject matter. We wish to acknowledge very highly the academic commitment of all contributors in putting this book together—in the process making this into a thrilling and very rewarding experience for both editors. It has been a great pleasure and satisfaction working with you all in this.

Theano S. Terkenli
Anne-Marie d'Hauteserre

THEANO S. TERKENLI

LANDSCAPES OF A NEW CULTURAL ECONOMY OF SPACE: AN INTRODUCTION

Christmas in Monaco: princely landscape to mesmerize visitors. Source: A.-M. d'Hauteserre

In the context of a fast changing world, forces of geographical transformation— "development" and "free market" capitalism, time-space compression, media and communication technology revolutions, "globalization", exploding patterns of networking and geographical flows, etc—acquire new facets, properties and directions, invariably reflected and imprinted upon the landscape. Change constitutes a much acknowledged reality that is neither novel nor unique to one geographical region over another. However, in the so-called "postmodern" world, it

1

T.S. Terkenli & A-M. d'Hauteserre (eds.), Landscapes of a New Cultural Economy of Space, 1-18.

seems to be acquiring a set of characteristics that invite investigation as to their distinctiveness, extent and nature, in terms of newly-emerging time-space-society contingencies. This basic realization constitutes the reason d'être of this volume.

Though this acknowledgment has varied enormously among scientists and lay-people, a broad range of ongoing changes in space and landscape is acknowledged to be emerging from all sides and sectors, and reflecting upon all, of postmodern life. Principally a First World phenomenon, these changes refer back to processes with distinctive geographical, historical and cultural articulations, some widely familiar, some even shockingly novel. They have been and continue to be much addressed, but inadequately spun together into frameworks of analysis and interpretation. Their various guises, aspects and manifestations in terms of landscape forms, functions and meanings/symbolisms call for a deeper engagement with them, taking in consideration the breadth of their occurrence and scope and the multitude of scales at which they materialize. They demand a more concerted, focused and systematic engagement with them, their provenance and repercussions. Towards this goal, the book rests on certain basic tenets, further developed and elaborated in the chapters that follow.

First, culture is central to the articulation of present-day socio-spatial transformation. Obviously, transformation does not occur in a vacuum. All aspects of life come into play in forming and shaping change as accounted for above, and introduced in this volume as processes of a "new cultural economy of space" (Terkenli, 2002). Secondly, such processes of a new cultural economy of space do not constitute a wholesale *new* reality; they are not everywhere, not new everywhere and not the whole story. They merely represent tendencies, whose outcome is far from clear and obvious. They are still evolving in multiple, complex ways, sometimes erupting in groundbreaking new realities, albeit in some kind of close connection to older structures of thought, power, practice and meaning. As such, thirdly, they are most evident in the landscape. Conscious or unconscious application and expression of such transformation in human contexts of life becomes most direct and discernible in the landscape, the most eloquent and "natural" geographical medium and product of such change in human life and activity. The projection onto the landscape—the interface of human-environment interrelationships—of contemporary change is another subject so far little explored and theorized in a systematic way by geographers.

On these premises, this collective effort seeks to contribute to theoretical advances, analytical approaches and applied studies in the broader inter(trans)-disciplinary field of contemporary research in landscape change. It seeks to bring together variable perspectives, insights and constructions pertaining to contemporary landscapes and landscape representations from different theoretical and methodological positions, as well as from diverse geographical and historical contexts, in order to elucidate and illustrate processes of cultural transformation, such as the ones described above. The overarching question is: how do these processes work in different geographical contexts and contribute to place and landscape creation? Perspective matters; the generation of scientific questions becomes all the more possible, fertile and proliferate, in the case where perspectives come together, exposing sources and meanings of questions posed.

THE CONTOURS OF THE NEW CULTURAL ECONOMY OF SPACE

Much discussion and speculation has engaged disciplinary and lay geographies in the subject of rampant "global" change occurring during the past few decades. It has acquired various guises and properties, instigated by various concerns and positions and has resulted in a plethora of—ill-judged or not—tenets, arguments and aphorisms about processes of contemporary spatial transformation. Aspects of such transformation have been addressed in terms of "time-space compression", "globalization", "information economy", "experience economy" and now "lifestyle economy", "transculturation", post-industrial society, etc., while its products have been described as "virtual geographies", "cyborgs" and landscapes of consumption, "Disneyfication", commodification, "placelessness", hybridity, "heterotopia", etc. Though many of these discourses admittedly lapse into universalizing hyperbole, it nonetheless refers to actual ongoing processes resulting into *new* forms of spatial organization constantly produced and reproduced at all geographical scales. Increasingly affecting and informing cultural landscapes of the Western, at least, world, these processes acquire the characteristics of what is termed here "a new cultural economy of space". They increasingly also affect the rest of the world, albeit very unevenly, as capital seeks ever more locations where to raise profits.

Spreading from the postmodern Western world, this unfolding global cultural economy of space is conceptualized as a cultural but still very much profit motivated, in the broader sense of the term, renegotiation of space. Technological change probably constitutes the most influential set of factors at the basis of this long-developing momentum of geographical transformation: specifically, the possibility of distanciation and reproduction in human-space interactions. The break from spatial exigencies, including not only transportation and communication inventions, but also the advent of mechanical reproduction of spatial forms and functions, typography, photography, video, digital reproduction and electronic technology, have all been leading to current cultural apprehensions, visions and constructs of space and landscape. According to the Dictionary of Human Geography, the ability to produce detailed, moving, three-dimensional environments is now reaching the point where these environments are becoming a significant supplement to the landscape around us, or even new kinds of landscapes (Johnston et al., 2000, p. 891).

Places and landscapes have always been organized on the basis of specific cultural economies of (time-)space. The much debated novelty of most of these forces, factors and processes of change notwithstanding, contemporary change is occurring at a much more rapid pace than in the past. It often materializes in new forms and shapes; it generates new mental, affective and symbolic schemata. Most importantly, however, it develops structures and functions of spatial organization that transcend previous sectoral interconnections around the globe, as in the markedly uneven functional integration of globally dispersed activities and networks.   Though present for at least several decades, these tendencies (internationalization, integration, networking, etc) are of a qualitatively different nature than in the past—to be further developed in the following exploration of the distinctive characteristics of the new cultural economy of space.

The term "cultural economy", as coined here in its broader sense, parallels the traditional usage of terms such as "political economy" or "cultural economy of contemporary history", by bringing together culture and the arrangement/mode of operation and/ or management of space (adopted after Webster's 1983 Ninth New Collegiate Dictionary definition of economy), or, in this case, local affairs, where the local and the household extend to the whole planet as home at the global scale. The adoption of the term implies the recognition of economy as a cultural site, as much as any other social domain—culture inherently affects and interweaves with economics.   This is not to privilege the binary culture-economics: economic and cultural,   as   well   as   other,   forces   have   always   been   functioning   together simultaneously. Aspects of culture, for instance, have been co-opted by capitalism for private or public economic gain.   Such interventions inadvertently embed themselves in the landscapes they create, as exchange values are split from use values.   Rather,   it   is   an   attempt   to   look   at   the   entire   nexus   of culture/economics/politics, to privilege a discursive understanding of place, re-establishing the intrinsic relatedness of the contents of place and landscape. Towards this goal, this collective work represents both an ontological and an epistemological argument, where culture becomes the central organizing principle of spatial change.

This "new (global) cultural economy of space", then, emphasizes a cultural negotiation and interpretation of newly-emerging spatial patterns, relationships and impacts; it constitutes more of a culture-centered approach of space rather than one exclusively centered on the uneven geography of costs and revenues. The relevance of a cultural understanding and interpretation of the changing geographical schemata of changing socio-economic relations becomes more obvious and instrumental in the case of the landscape than any other spatial unit—one of the basic positions adopted in this collection of essays.

Although the cultural constitution, articulation and materialization of current spatial transformation are upheld in most cases, not all authors use the term cultural economy of space exactly in the sense specified above. Precisely due to its all-inclusive definition, the term is conducive to free adaptation by investigators of the multitude of phenomena it encompasses.   Such variegated usage of the term is especially wishful and welcome here, in that it promotes a multitude of approaches and means to the study of contemporary landscape change. It reinforces alternative structures of understanding, induces proliferate geographical imaginaries and, in this context of acknowledgement, strives to incorporate differentiation. What unites the efforts of all those contributing to this volume, on the other hand, is the quest for commonalities,   the   search   for   some   sort   of   sameness,   in   ongoing,   spatially differentiated change. It must be emphasized, nevertheless, that change of whatever sort is not geographically uniform and, hence, apparently universalizing schemata such as the one proposed here should be conceptualized and applied contextually only as a complex of ongoing interrelated processes, highly uneven in time and space—and not as end results.

Social scientists and scholars of several other provenances and affiliations have long been negotiating processes of spatial change. The outcome of this substantial and growing body of work is manifested in a plethora of research conclusions in various fields of scientific knowledge. In terms of new landscape forms and shapes, striking novel apparitions have been recognized, produced and reproduced in architectural negotiations of space, urban and regional planning and new forms and technological applications in the organization of rural spatial systems. New economic structures and networks, resulting, for example, from processes of multinational-corporation restructuring, have created new geographies of power and political might. Alternative lifestyles and values, stemming from the rapprochement of cultures at all geographical scales, have been creating societies of consumers, "global citizens", nomadic elites, cyborgs, etc. Exploding and imploding patterns of recreation and tourism are altering the face of the world and of the landscape. New structures, values and processes in recreation and public life are increasingly modifying the landscape, often leading to irreversible change in the pre-existing landscape, as in the case of thematic parks, golf courses and shopping malls. On the other hand, many of the disenfranchised poor have traveled to improve their living conditions: immigration, expatriation and repatriation have been creating new hybridities in metropolitan centres, but also often brought back into their own home culture, challenging, engaging and reformulating local realities. Meanwhile, in this era of transnationality and even postnationality, geographical scales intermingle and interweave in continually evolving new ways—i.e. the translocal and the transregional levels—serving new types of functions and processes made possible through the ongoing technological revolution in networking and communication.

In an attempt to navigate through the contours of the different facets of this new cultural economy of space, some of its most elemental characteristics may be summed up here, to be reviewed again at the closing of this volume: a) new collective experiences/sense of place that increasingly transcend geographical barriers of distance and of place and create new geographies of time-space; b) a growing de-differentiation in space between private and public spheres of everyday life, rearticulating complex relationships between the personal and the social; c) de-segregation of the realm of leisure from the realms of home and work life; d) changing geographical schemata of changing socio-economic relations at variable geographical scales; e) the rapid and overarching exchange and communication of symbolic goods (flows of money, ideas, information, images, etc), f) through variable processes of networking and globalization where visual media predominate over textual media (Terkenli, 2002).

PROCESSES OF THE NEW CULTURAL ECONOMY OF SPACE IN THE LANDSCAPE

If tradition may actually have a short history (Hobsbawm & Ranger, 1992), novelty may conversely have an amazingly long history, indeed. The mechanisms of change are simply always at work. The revolutionary character of technology in the past few decades has presented a growing segment of humanity with the potential of interconnection into a series of networks. Colonialism and the modernization project has triggered landscape homogenization, a project since going strong, and bearing a long line of diverse spatial repercussions. It may be, for instance, that the attempts by urban planners and architects to plan and homogenize urban space have resulted precisely in the opposite: increasing formlessness and heterogeneity in the urban landscape, strongly differentiated on ever more levels.

Until recently, landscape identity used to be articulated in the context of a particular socio-economic system embracing and expressing the local dynamic of land and life. The increasing porousness of temporal and spatial barriers and the explosion in movement and interconnectivity in the Western world wrought great fragmentation, differentiation, conformity and/or complexity both between and within what formerly used to be more distinctive and homogeneous landscapes. We propose to address processes of this new cultural economy of space with the aid of the tentative terms "enworldment", "unworldment", "deworldment" and "transworldment" (Terkenli, 2002), serving to elucidate and organize current trends in the transformation of existing geographical schemata. These terms purport to apply to all levels of geographical analysis, such as cultures or landscapes, life spheres (work, home, leisure), lifeworld realms (public, private), social groupings (on the basis of class, race, ethnicity, and so on) or other frameworks of analysis. They may not necessarily or directly be space-based; they may adhere to the realms of the real or the virtual, the imaginary or the artificial, the extraordinary or the familiar etc. Enworldment, unworldment, deworldment and transworldment do not represent a continuum, but rather sets of processes operating more or less simultaneously, irrespective of the order presented here, though some of them tend to be initiated or reinforced by some more than by other ones of the above processes. Admittedly, the analytical task of disentangling one such set of transformative forces from another, as well as of differentiating their geographical impacts, remains extremely complex and challenging—if ever possible.

In particular, enworldment processes refer to the breakdown of barriers and boundaries between previously existing worlds, on the basis of any geographical or substantial analytical schema. The contemporary blurring and fusion of conceptual and actual categories, either spatial or substantial, in which the human world may be compartmentalized, signal and usher in variable spatial transformation. This transformation is reinforced to the degree to which old spatial and social schemata are dismantled or altered and new ones created. The dismantling of old and established socio-spatial structures signifies processes of unworldment, while growing disassociation of these new schemata from geographical location and unique place characteristics implies processes of deworldment. Processes of unworldment, operating through globalizing forces and homogenization tendencies, signal the gradual loss of place and landscape identity, for instance in terms of authenticity or in terms of a sense of place. Processes of deworldment may be framed in new sets of rules that defy common existing practices and conceptualizations of space and may be accompanied by ground-breaking trends.

Such transformation forces invariably take on increasingly global dimensions (transworldment), although their manifestation obviously varies over space, time and social context (Terkenli, 2002).

The ending «worldments» has been selected for our purposes, in order to expose and emphasize the broad and increasingly globalized scope of ongoing change through processes of the new cultural economy of space. The coinage of these terms aims at the creation of a more geographical terminology that addresses contemporary spatial change. The term globalization appears too generalized, its meaning too fuzzy and highly contested for our purposes; lacking in nuance and detail as to geographical scale and dynamics of change. Moreover, it is suggested that in their description of very distinctive spatial products and dynamic of change, processes of the new cultural economy of space are especially suitable to the study of landscape. Whether change is postulated as globalization, commodification, development or Disneyfication, its landscape products are all an integral and obtrusive part of late twentieth- and early twenty-first-century modernity and postmodernity. As illustrated further down, these processes of change are much more overtly and explicitly represented in the landscape than in other geographical units of analysis. On the basis of its easy and ready accessibility, imageability and representability, as we shall see throughout this volume, landscape constitutes a most significant geographical medium in the analysis of processes of the new cultural economy of space.

If place is a spatio-temporal intersection of «a particular constellation of [human] relations» (Massey, 1993, p. 66), then landscape is its image, as reflected in the relationship of the human being with a specific geographical setting. Landscape is thus, at the outset of our discussion, conceptualized as a visible expression of the humanized environment. In the landscape, the visual/ material, the experiential / functional and the symbolic/ cognitive (otherwise, form, function and meaning) come together, rendering landscape a valuable means and tool of geographical analysis. Of the three highly interrelated and interactive sets of landscape properties, however, in the Western world, landscape has been traditionally defined on the basis of its imageability and relational character (observer-based definition) (Terkenli, 2001). Accordingly, what is upheld here is not an essentialist notion of the landscape, but rather a culturally ambivalent, socially constructed and historically specific notion of the landscape that invites multiple and fluid interpretations, among which certain ones have historically prevailed.

Landscape, though never a self-evident object in Geography, has been one of its most resilient terms, whose «theoretical framework always structured its interpretation; it was an analytic concept which afforded objective understanding» (Rose, 1996, p. 342). Landscape is shaped by both biophysical laws and cultural rules, interpreted and applied to the land through (inter-) personal and (cross-) cultural strategies (Jackson, 1984; Naveh & Lieberman, 1994; Rackham & Moody, 1996). Thus, its articulation has depended on both objective and subjective ways of understanding, fully encoding the essential «betweenness of place» (Entrikin, 1991), otherwise conceptualized as «the duplicity of landscape imagery», going beyond the single vantage point of a spectator, to «work up an idea of human geography, a view of country life and regional character» (Turner in Daniels & Cosgrove 1988, p. 7). Consequently, landscape has been viewed not just as a material expression of a particular relationship between land and humans, observable in the field through an objective gaze, but rather as a way of seeing, «a cultural image, a pictorial way of representing, structuring or symbolizing surroundings» (Daniels & Cosgrove, 1988, p. 1).

Implicit in these definitions, as well as inherent in the philosophy of our approach, are various perspectives to specific geographical contexts, as well as variable social and environmental meanings and relations as understood and inscribed in the landscape (Meinig, 1979; Daniels & Cosgrove, 1988; Cosgrove, 1998; Aitchison et al., 2000). Thus, generally speaking, landscape analysis is highly contingent on time, space and social context and necessitates widely varying goal-specific analytical approaches: a challenging, inherently multi-(trans-)disciplinary task that addresses a wide variety of landscape functions, forms and meanings (Jakle, 1987; Taylor, 1994; Buttimer, 1998; Stefanou, 2000; Tress et al, 2001). Consequently, as a focus of research, the landscape requires contextual interpretation and cannot be detached from questions of positionality and from its historical and sociocultural context—its relationship with an observer. It is proposed that the chapters in this volume serve and illustrate this conviction well; in fact, this conviction serves as the main objective of this collective effort.

ENWORLDMENT, UNWORLDMENT, DEWORLDMENT, TRANSWORLDMENT

Processes of enworldment (Weaver, 2000) selectively compress and condense geographically distinct versions of the world into single landscapes, while simulating a multitude of various other landscapes, striving to create competitive poles of consumption, attraction and spectacle. Enworldment signals the geographical transferability and encompassing of previously existing worlds in one: a direct outcome of time-space compression and the blurring of boundaries in space and time and among spheres of the human lifeworld (i.e. between home, work and leisure). These processes also express the breakdown in the distinction between culture and nature, a quintessential ontological division in geography, as exemplified in computer-simulated wilderness landscapes and zoos. The repercussions of processes of enworldment are inscribed in the landscape as a complex and highly attractive mix of old and new, familiar and different, all produced and consumed in situ provided that "it sells". The resulting landscapes come in various forms and guises—often they represent a simple impression or a certain sense of landscape. As one part of an urban landscape acquires a strong profile as front, for instance, inevitably a "back" appears, just as unused areas left over by planning, situated outside its field of action (Nielsen, 2002): "superfluous" landscapes of various sorts thus appear in urban contexts as the natural outcome of efforts towards homogenization, standardization, commercialization, aesthetization or other such universalizing spatial interventions.

Processes of enworldment concentrate all possible amenities, attractions and privileges in seductive landscapes, the ultimate example of which may lie in Las Vegas. These landscapes tend to be characterized by the breakdown of internal barriers between lifeworld spheres and opened up, even in their most private sections, to the all-consuming gaze or dollar/EURO/Yen etc. Central to processes of enworldment are obviously the increasing extent and impacts of commodification. The investment of a landscape or landscape myth with high exchange value for commercial purposes follows a trajectory of commodification that takes on several attributes and unfolds through variable processes of landscape objectification, aesthetization, pictorialization, all in an effort to impress and invite investors or consumers. Culture and landscape are thus staged, sacralized and commodified for purposes of satisfying contemporary unequivocal or homogenized tastes and omnivorous desire for culture and landscape consumption. The consumption of *signs of* commodities, as well as of landscapes, has been gaining ground over commodity consumption itself; in the process, place, cultural and personal identities dissolve and are recreated (Miller et al., 1998; Jackson, 1999). Unfolding in a broad range of contexts of different worlds, processes of enworldment create all sorts of new landscapes characterized by unequal human relationships between the intersecting sides. These inequalities refer mainly to power and economic disparities (Harvey, 1989), but may be equally articulated on the basis of gender, ethnicity, cultural system, race, sexual orientation or other axes of personal or collective identification (Massey, 1994; Dear & Flusty, 2002). Their intersection and co-existence opens up a wide range of new possibilities, ambiguities, contradictions and tensions in landscape commodification, attraction and consumption.

Processes of enworldment invariably initiate and often bring about an inevitable loss of pre-existing place and landscape identity through processes of unworldment. Unworldment processes dissolve geographical particularity, landscape distinctiveness/ identity and place attachment (i.e. a sense of home), as formulated thus far. They signify the collapse of geographical and substantial categorical distinctions (Sack, 1980) and signal the undoing of "known" landscape geographies. Their outcomes are often described in terms of «inauthenticity» and «placelessness» or in terms of a loss of a geographical sense of place. The focus on the loss of place identity in terms of "inauthenticity" represents a top-down perspective on spatial change, uniqueness and distinctiveness, whereas a focus on placelessness represents a bottom-up perspective on the sense of place, i.e. as home for its inhabitants. These observations on place obviously also extend to landscape, as a personal or cultural image of a place.

Inauthenticity, a concept much discussed in tourism studies, for example, describes changes that alter the character of place and landscape, commonly occurring at mass tourist destinations or through the prolonged presence of outside influences at a locality (MacCannell, 1973; Salamone, 1997; Wang, 1999; Taylor, 2001). In numerous locations around the world, "traditional" villages readily spring forth at the lure of tourism. Sorkin describes such processes caused by forces of globalization as processes of dissipation of all stable relations to local physical and cultural geography, the loosening of ties to any specific space (1992, p. xiii-xiv). Placelessness, on the other hand, addresses the loss of the very essence of place and of landscape: the value and meaning invested in a geographical location that distinguishes it from abstract, undifferentiated, «objective» space. This concept refers not only to place and landscape character loss, but also to the total loss of a sense of meaning and use value of place and landscape (Relph, 1976). Perhaps the most significant impact of unworldment processes is the loss of the sense of place and landscape as home for its inhabitants. The distinctiveness of home as a type of place is established through a) steady, cyclical investment of meaning in a context that b) through some measure of control or claim, a person or a group of people identifies with (Terkenli, 1995). The advent of mass tourism and resulting processes of unworldment disrupt both of these distinctive characteristics of home. The disruption of whole towns and villages by the unplanned presence and unregulated functions of tourism in communities that are unprepared to host large numbers of visitors may lead to a partial or wholesale renegotiation of social relations, local ways of life and their cyclical rhythms as these are related to both private lifeworlds and public systems.

Processes of unworldment often lapse into processes of deworldment, with the creation of fictitious, commercialized, ephemeral, disposable, staged, «inauthentic» landscapes and worlds of recyclable and expendable illusion. They produce landscapes autonomous from "reality" and prone to host an infinite series of possible historical and geographical self references. Such transgressive geographical juxtapositions are especially common in the fluid contemporary context of the so-called information economy of the new electronic age (Castells, 1996). Deworldment may also come about as a direct outcome of enworldment, as well as of processes of touristification, commercialization and cultural banalization. These «Disneyfied» landscapes are often described as controlled microcosms of paid-for public activities, normative, sterile, proper, self-referent, predictable, clean, unpolluted etc, where the subject is simultaneously actor and spectator. They are characterized by place and landscape deconstruction and redefinition. The spatial products of these forces at play may resemble skewed, incongruous or surreal landscapes, such as Foucault's «heterotopias» and Baudrillard's «hyperspace», or simply contrived spatial entities where the artificial, the virtual and the staged imitate the «real» or the natural, and even seek to surpass them in terms of originality. They include Andy Warhol "high art", pop star Luciano Pavarotti, and authentic Mondrians in Las Vegas casino halls. In general, (o)u-topias ("no-places") liberate from the constraints of space and place, while their surreal counterparts, "Disneylands", effectuate escape into the world of fantasy and life as a spectacle. Utopias, as well as heterotopias ("places of otherness"), only exist in contrast and because of *topoi* ("places"), through an ongoing ambivalent and complex dialectic of spatial continuity and change.

Processes of deworldment may also lead to place and landscape devolution and consequent human decentering in their midst, or to new landscapes of the collective imaginary. Psychological and symbolic impacts and consequences of processes of deworldment may vary greatly with regard to modern and postmodern constructs of seduction and attraction and expressions of desire. The subject, in a state of dislocation, seems to be constantly on the move in non-space: at an extreme stage, (s)he lives a series of temporary spatial nonengagements, bypassing the local, yet always well connected around the world. Subject and urban space/ landscape are increasingly perceived as products of new communication technologies, developing more direct interrelationships. This human-space interrelationship sometimes acquires tense, unsettling and contradictory dimensions. The subject, as a cyborg, expands in the urban body in his/ her constant interaction with the urban landscape, by combining, in ways arbitrary and fragmented, diverse and diffuse socio-spatial phenomena, activities, processes and relationships—the vital organs of the city (Haraway, 1991). This is a two-way relationship, naturally: the urban landscape is, in turn, inscribed in the subject which mimics it and refers to it in a symbiotic relationship where cyborg and space lose their autonomy.

Shopping malls combining retail and leisure elements now constitute the most significant recreation landscapes for middle-class America, drawing large numbers of visitors. Obviously, leisure shopping and retailing are nothing new; what has changed, however, is the range of leisure-retail environments now available (Shaw & Williams, 1994, p. 206). Shaw and Williams, in specific, point out that «such developments blur the distinctions between so-called primary and secondary leisure products, as tourists become increasingly attracted to these stage-managed shopping experiences» (1994, p. 207). One significant emergent characteristic of these new and rapidly diversifying types of landscapes, then, is their wholesale, maximal orientation towards the consumer's desires and fantasies, in saccharine thematic/ iconographic stereotypes, tending to a nostalgic resurgence of harmless innocence and catering to the demand for easily digested body/ mind/ spirit experiences. This argument specifically holds in the case of urban forms of tourism, where leisure— especially as this relates to the arts, food and drink consumption and leisure shoppin — becomes a central element of the so-called new symbolic urban economy (Deffner, 1999, p.119).

In the same line of thought, while theme parks may be viewed as landscapes of family entertainment, privately-owned themed environments in general target more diverse tastes and demands. The striking growth of theme parks (Williams, 1998) as the quintessential postmodern illustrations of place and landscape attraction, may represent perhaps the most eloquent yet example of processes of deworldment. These new types of Never-Never Lands are characterized by a breakdown of distinct rules of spatio-temporal division and social practice, such as between work, home and leisure or between high art and popular entertainment. The development of themed environments, however, generally goes beyond this de-differentiation of spaces, functions, styles and symbolisms and the deliberate blurring of the real with the artificial and the imaginary. It rests on the effectiveness of the idea of «invented» landscapes and places and aims at creating contemporary wonderlands of selective nostalgia and pseudo-idealistic visionary. When placed in a fully themed environment, the subject is given an already-interpreted landscape, a ready-made world (Rodaway, 1995, p. 262). Baudrillard, in this case, talks of a subject that turns into «a kind of ecstatic object, a continuous circulation of signs, a replicating and metamorphosing body, driven by hedonistic conformity to the order of simulation--represented in part by the images of the mass media, both advertising and programmes» (Rodaway, 1995, p. 264). This point brings us to the last facet of spatial change instigated by processes of the new cultural economy of space.

Processes of transworldment complete and transcend the dissolution of cultural boundaries characteristic of the postmodern age; they manifest in the constant reproduction and widespread projection of landscapes and geographies through actual, virtual or imaginary connections and flows. As the world becomes increasingly interconnected through all sorts of networking, it is argued that certain divisions are harder to sustain, leading to the creation of "third-space" and landscapes of hybridity. These arguments suggest the need to think of space and landscape in particular ways, such as in terms of flows and connections rather than of localized constructs and activities. The vast proliferation of media plays a large role in the ways that the dissemination of images, texts and sounds create new types of landscapes in our information economy and network society: mediated, electronic, ephemeral, standardized, detached and instantaneous. Transworldment processes are reinforced and accelerated by the ongoing revolution in the generation and transmission of all sorts of information and the emergence of a knowledge theory of value, calling for new cultural apprehensions of space and landscape— within or beyond the realm of the possible. In turn, transworldment processes provide the impetus for the reinforcement and regeneration of processes of enworldment, unworldment and deworldment—not necessarily in this order. Each one of these sets of processes usually unfolds conjointly with other sets of processes of the new cultural economy of space. Their manifestations and impacts are, consequently, difficult to disentangle and identify with one or another category, while some of these new spaces/landscapes are perhaps not yet recognizable. Moreover, the workings of these processes necessarily create residue: in the interstices of the planned, of the used and managed urban space, in terms both material and immaterial, spring forth radically new and highly differentiated possibilities, opportunities and outcomes (see chapter 3 by Lange in this volume).

Visual media predominate over textual media in the context of this new cultural economy of space. In this context, the role of the icon, as well as of its accompanying text, according to Jacques Ellul, is «pre-propagandistic» (Papaioannou 1999, p. 114); that is, it is preparative of the grounds on which the intended message will be transmitted and more or less subtly suggestive of intended stereotypes or disruptive of existing ones. Daniels and Cosgrove argue for the centrality of the iconographic method to cultural inquiry: «[E]arlier and less commercial cultures may sustain more stable symbolic codes but every culture weaves its world out of image and symbol», beyond objective, attractive and orderly views of the world as dazzling pernicious distorted delusions (1988, p. 7-8). Iconography, the theoretical and historical study of symbolic imagery, equally refers, according to Panofsky, to built as well as to painted forms (Daniels & Cosgrove, 1988, p. 1-3) and by extension to landscapes as well as to their images.

Thus, it becomes the ideal means for the dissemination of messages, forms and symbolisms through processes of transworldment.

Here the distinction between two visual forms of communication must also be emphasized, that between the sight (seeing) and the spectacle (gazing). While sights are predominantly experienced with the aid of the sense of vision and apprehended as solidity and temporal transcendence, the experience of spectacles is both differentiated and temporally bounded (MacCannell, 1992). Spectacles incorporate time and space management, producing landscapes frozen in time-space and offer the satisfaction of acquiescence, familiarity, intimacy, acknowledgment, control, entertainment and comfort. The link between spectacular action and emotional response is direct (MacCannell, 1992), intensified by the power and impact of instant visual transmission.

In their iconic or virtual form, spectacles become simulated landscapes and pseudo-events. It can be argued, of course, that virtual reality has already existed for a long time, as in eighteenth century "picturesque" landscapes, etc. Humans have often been able to engage best with what is not "out there", anyway. The novelty of such environments in the present age, however, lies instead in their nature, scale and geography (Crang et al., 1999)–characteristics that will be much discussed throughout this book. More significantly, such trends and associated developments cut across much of the more traditional landscape typologies, forming new types of landscapes of the new cultural economy of space. One outcome is that landscape is no longer nature, not even an expression of place identity or cultural image. It is timeless, spaceless, cultureless, nationless. It is commonly a product, produced for the purpose of wholesale consumption in any and all of its dimensions: visual/aesthetic, functional/experiential and symbolic. Beyond such widely upheld aphorisms, however, actual world circumstances are much more complex, geographically and historically differentiated. Rather, if complexity had always applied to the human-environment relationship, today it seems to be more technologically sophisticated, developed and intensified by processes of enworldment, unworldment, deworldment and transworldment, apparently creating new ways of relating to the landscape that are much more fluid, complex, surreal and a-geographical than in the past. These and other forms and consequences of spatial transformation are explored in the chapters of the volume at hand.

THE STRUCTURE OF THE BOOK

This collective volume is structured into six parts, largely corresponding to the four sets of processes of the new cultural economy of space: enworldment, unworldment, deworldment and transworldment. These themes are, however, carried throughout the whole book in the individual chapters, from one part to another, in no particular ordering, just as the processes of the new cultural economy of space, interwoven, intermingled and overlapping, reflect "real world" situations. What follows is a brief introduction into the individual chapters. There is no way in which we can do justice to all aspects of these contributions; indeed there is a danger of oversimplifying their positions. "Presentation is always concerned with seizing power or deferring to it; one needs to recognize this, not only when making decisions about how to present one's work, but also when trying to work out how to understand what is laid before one" (Shurmer-Smith, 2002, p. 7). Suffice it to acknowledge here the range of interrelating theoretical positions and methodologies, themes and scales, as adjusted to the context at hand and the circumstances of the investigation. One may detect a playful attitude flowing in and out of some of the texts at hand, alternating with a more exacting and admonishing one to an alarmed or a polemical one. In any case, most of the empirical case studies spring forth out of the authors' personal geographies, the contexts of their own everyday lives. Thus, they carry great potency and dynamism in their relevance to the unifying themes and goals of this book and begin to illustrate the omnipresence of processes of the new cultural economy of space in our everyday life contexts in the Western world.

Chapter 1, authored by David Crouch, opens the scene by focusing upon the mechanisms through which the individual and the human body are engaged and figure in the new cultural economy of space. Its contribution lies in the re-attachment of the human component of the world to the "mediated cultural context of the destabilized landscape of an ongoing spacing of the world". It proceeds from a presentation of the theoretical background of the new cultural economies; in this way, this chapter substantially complements the introduction of the book. The author proceeds to counterbalance the workings of globalization through the making of alternative economies and resistance and by individual productions and circulation of meanings in life-practices and life exchanges. He argues for more active and negotiative ways in engaging the world and its meanings from the bottom-up. Bastian Lange's essay (chapter 2) provides an example of new spatial strategies resulting from processes of enworldment in the urban landscape of Berlin, in the case of the "Culturepreneurs", working at the economic interface functions of the city. Culturepreneurs function under the constraints of social hardship brought about by processes of the new cultural economy of space, at the interstices of different forms of institutional integration. They develop spatial placing practices and movement performativity patterns which stand in complementary or ambivalent relationship to urban spatial organization strategies: a new type of hybrid cultural, as well as entrepreneurial, agents who produce and perform new models of urbanity and "microglobalizations".

In chapter 3, Denis Cosgrove applies unworldment processes to contemporary urban space, while, at the same time, contests the theoretical implications of the term unworldment, as well as the novelty of such transformative processes in the landscape. By juxtaposing two "superficially very different landscapes", he reveals significant similarities in their geography and evolution, which, he contends more and more regions today share. Through an in-depth investigation of these two "paradigm postmodern landscapes", L.A. and the Veneto, the author illustrates the historical evolution of the spatial model of leisured life and communication technologies in the urban landscape, increasingly relevant to the construction of postmodern space. Similarly, chapter 4, by Andrew Sluyter, contributes to the substantiation of unworldment processes, as applied to the integration of culture theory into development theory and political intervention in the case of Mexico. The author achieves his goal by leaning on Mary Louise Pratt's focus on von Humboldts's travels in and subsequent writings on Mexico. For this purpose, he emphasizes his use of landscape, as a geographical entity fully incorporating the spectrum of biophysical and social processes, as he asserts in his exposure of the linkages between unworldment and "undevelopment". The chapter thus takes a step towards understanding and explaining the processes through which colonialism and postcolonialism have unworlded procolonial worlds, aiming towards an application of such understanding to socially and environmentally-grounded development in today's postcolonial world.

In chapter 5, Katriina Soini, Hannes Palang and Kadri Semm address the long-standing issue of bringing peripheries to the same level as more privileged locations in terms of development and global networking. The authors recount rural change, decline and restructuring in a comparative study of landscape transformation in Finland and Estonia. Their approach of processes of deworldment balances other case studies and perspectives in this volume, by focusing on the rural landscape and by grappling with microeconomic and community issues at the local scale. The chapter analyses complex tensions in the landscape as a result of changing socio-economic formations, in terms of insider-outsider relationships, place making and place loss, depopulation and cultural revitalization, landscape continuities and breaks and dismantles the notion of "traditional" landscapes, relegating them to the realm of nostalgia. In the same line of thought, d'Hauteserre, in chapter 6, places her critical investigation of tropical landscapes of tourism in the context of the new symbolic economy of space, on the basis of the cultural construction of tourism as an economic industry. As imaginary constructs of unworldment and deworldment processes, tourism landscapes are shown to be occupying the periphery of the "real world", especially in the "exotic" geographical periphery of the Tropics, colonized by processes of the new cultural economy of space from the first world. The chapter probes into the ways that discourses of third world landscape production and reproduction are developed, disseminated, apprehended and invoked, as well as differentially vested with meaning, depending on the side that does the landscape reading and interpretation. In so doing, it subverts notions underlying tourism development in the third world, such as notions of insiders-outsiders, time-space continuity, etc.

In the fourth part, Kenneth Olwig's essay (chapter 7), following de Certeau's work on space and place, turns our attention to the definition of place and landscape and to their structural antithesis: the vanishing of place and its elimination to nothingness. By drawing on the assault on the World Trade Center's global symbolism as a monument of imperial stature and its elimination to Ground Zero, this essay builds the significance of "an absence that is presence", or of the importance of zero to the understanding of place, landscape and the world. The essay illustrates other forms of transgressive practice leading to the emptying out of spaces and landscapes and to their stringing together anew with meaning. Its case studies (also Place de la Concorde, Paris and Rɛdhuspladsen, Copenhagen) thus contain the seeds of "reworldment" (see conclusions of this volume) through an in-

depth articulation of the mechanisms of the emergence of new forms of place/
landscape. Another novel spatial case of transworldment is explored by Jussi
Jauhiainen in chapter 8, in technologised landscapes attaining global symbolism
through the ongoing revolution in information and communication technologies
(ICT's). This chapter weaves its discussion on the nature of changes brought to
contemporary societies and everyday life through ICT's and their transformative
spread into the landscape, in parallel with a self-exploring journey into newly-
emerging forms of human-environment interaction through processes of the new
cultural economy of space. The author constructs the case of the landscape of the
town of Oulu—the Silicon Valley of Finland—with the aid of Zeitdiagnose, varied
empirical material and personal observation, through a dynamic subjective synthesis
of a newly-emerging type of landscape. In it, he apprehends and contests binaries
on the basis of which landscapes are constructed, such as local-global, past-future,
subject-landscape etc. Finally, the chapter by Martine Geronimi (chapter 9) also
places its negotiation of landscapes of tourism, in the context of an emerging
symbolic economy. By identifying processes of transworldment in Vieux-Qubec
(Quebec City), the author examines how French imageries of its landscape are
electronically disseminated and reappropriated by various parties at stake.
Following a theoretical investigation of the context and concept of symbolic
economy, she demonstrates how this World Heritage tourist landscape embodies and
reveals processes of the new cultural economy of space, indicative of "a new
culture-centered approach to space". Thereby, the Vieux Qubec landscape becomes
a symbolic good and high-end commodity of Quebec City's French past and
identity, ironically appropriated by the Anglophile side (promoters, developers).

The unifying theme, as well as the prospective contribution of this collective
effort, then, lies in the exploration of these developing forces and characteristics of
the new cultural economy of space in the contemporary cultural landscape. The
primary objective of bringing together geographical perspectives from various
subdisciplinary fields is to examine and discuss ways in which the complexities of
this newly emerging cultural economy of space are applied to various sorts of
landscapes, i.e. landscapes of everyday life, landscapes of tourism and recreation,
postcolonial and hybrid landscapes, landscapes of economic production, landscapes
of the street and of public life, "national" landscapes and so on. The subject of our
discussion is the transformation and liberation of landscape from certain older
constraints; its implications lie in the challenge to our conventional geographical/
landscape ontologies and epistemologies. A further goal is the attempt to tie these
approaches together in an inter(trans)disciplinary context for future landscape
research. Our intentions are to create more space for the development of landscape
discourse that accommodates both theoretical and empirical findings, as well as
methodological issues and practical applications pertaining to the contemporary
landscape, in order to identify trends, structures, technologies and practices defining
and articulating this new cultural economy of space. It is hoped that this endeavor
will generate many more questions and areas of inquiry pointing to new directions
currently developing in the study of landscape than the questions on the basis of
which this task was undertaken here in the first place. Our ultimate aim rests in the

contribution towards an ongoing exploration of tensions over the uses and meanings of space and landscape; over the creation, dissolution and metamorphosis of space; and over the interweaving of landscape scales, boundaries, and characteristics.

REFERENCES

Aitchison, C., MacLeod, N. E., and Shaw, S. J. (2000). *Leisure and tourism landscapes: social and cultural geographies*. London: Routledge.

Buttimer, A. (1998). Landscape and life: appropriate scales for sustainable development. *Irish Geography* 31 (1), pp. 1-33.

Castells, M. (1996). *The Rise of the network society, vol. 1, the information age: economy, society and culture*. Oxford: Blackwell.

Cosgrove, D. (1998). Cultural Landscapes. In Tim Unwin (Ed.), *A European geography*, pp. 65-81. London:Longman.

Crang, M., Crang, P. and May J. (Eds.) (1999). *Virtual geographies: bodies, space and relations*. London: Routledge.

Cresswell, T. (1996). *In place/out of place: geography, ideology and transgression*. Minneapolis: University of Minnesota Press.

Daniels, S. and Cosgrove, D. (1988). Introduction: the iconography of landscape. In Denis Cosgrove and Stephen Daniels (Eds.), *The iconography of landscape*, pp. 43-82. Cambridge: Cambridge University Press.

Dear, M. J. and Flusty, S. (Eds.) (2002). *The spaces of postmodernity: readings in human geography*. Oxford: Blackwell.

Deffner, A. (1999). Cultural tourism and leisure activities: their impact on city functions (in Greek). In Dimitris Economou and Giannis Petrakos (Eds.) *The development of Greek cities: interdisciplinary perspectives in urban analysis and policy* (in Greek), pp. 117-156. Volos: Thessaly University Press and Gutenberg.

Entrikin, J. N. (1991). *The betweenness of place: towards a geography of modernity*. Baltimore: The Johns Hopkins University Press.

Haraway, D. (1991). A cyborg manifesto: science, technology, and socialist-feminism in the late twentieth century. In *Simians, cyborgs and women: the reinvention of nature*. New York: Routledge, pp. 150-1, 162-3; http://www.stanford.edu/dept/HPS/Haraway/DyborgManifesto.html.

Harvey, D. (1989). *The Condition of postmodernity*. Oxford: Blackwell.

Hobsbawm, E. & T. Ranger (Eds.) 1992. *The invention of tradition*. Cambridge: Cambridge University Press.

Jackson, J. B. (1984). *Discovering the vernacular landscape*. Yale University Press, New Haven.

Jackson, P. (1999). Commodity culture: the traffic in things. *Transactions IBG*, NS 24, pp. 95-108.

Jakle, J. A. (1987). *The visual elements of landscape*. Amherst: The University of Massachusetts Press.

Johnston, R. J., Gregory, D., Pratt, G., and Watts, M. (Eds.) (2000). *Dictionary of human geography, 4th Edition*. Oxford: Blackwell.

Kooijman, D. (2000). «Machines and Theatres» for place-making; recent trends in consumption space. Paper presented at the conference *Habitus 2000: A Sense of Place*, Perth, Australia, Sept 5-9.

Lofgren, O. (1999). *On holiday: a history of vacationing*. Berkeley: University of California Press.

Lowenthal, D. (1961). Geography, experience and imagination: towards a geographical epistemology. *Annals of the Association of AmericanGeographers*, 51, pp. 241-260.

Loyd, B. (1982). Women, home and status. In James Duncan (Ed.), *Housing and identity: cross-cultural perspectives*, pp. 181-197. New York: Holmes and Meier Publishers, Inc.

MacCannell, D. (1992). *Empty meeting grounds: the tourist papers*. London: Routledge.

MacCannell, D. (1973). Staged authenticity: arrangements of social space in tourist settings. *American Sociological Review*, 79, pp. 589-603.

Massey, D. (1994). *Space, place and gender*. Cambridge: Polity Press.

Massey, D. (1993). Power geometry and progressive sense of place. In J. Bird, B. Curtis, T. Putnam, G. Robertson and L. Tickner (Eds.), *Mapping the futures: local cultures, global change*, pp. 59-69. London: Routledge.

Meinig, D. (Ed.) (1979). *The interpretation of ordinary landscapes: geographical essays.* Oxford: Oxford University Press.

Miller, D., Jackson, P., Thrift, N. J., Holbrook, B. and Rowlands, M. 1998. *Shopping, place and identity.* London: Routledge.

Naveh, Z. and Leiberman, A. (1994). *Landscape ecology: theory and application*, 2nd Edition. Berlin: Springer.

Nielsen, T. (2002). The return of the excessive: superfluous landscapes. *Space and Culture*, 5 (1), pp. 53-62.

Papaioannou, H. (1999). Traversing Cyprus with a camera: wavering between beautiful and wild nature (in Greek). *Gramma: cultures of vision/ visions of culture,* 7, pp. 105-120.

Rackham, O. and Moody, J. (1996). *The making of the Cretan landscape*. Manchester: Manchester University Press.

Relph, E. (1976). *Place and placelessness*. London: Pion Limited.

Rodaway, P. (1995). Exploring the subject in hyper-reality. In Steve Pile and Nigel Thrift (Eds.), *Mapping the Subject:Geographies of Cultural Transformation*, pp. 241-266. London: Routledge.

Rose, G. (1996). Geography as the science of observation: the landscape, the gaze and masculinity. In John Agnew, David N. Livingstone and Alisdair Rogers (Eds.), *Human geography: an essential anthology*, pp. 341-350. Oxford: Blackwell.

Sack, R. D. (1980). *Conceptions of space in social thought: a geographical perspective.* Minneapolis: University of Minnesota Press.

Salamone, F. A. (1997). Authenticity in tourism: the San Angel inns. *Annals of Tourism Research* 24 (2), pp. 305-321.

Shaw, G. and Williams, A. M. (1994). *Critical issues in tourism: a geographical perspective.* Oxford: Blackwell.

Shurmer-Smith, P. (2002). Introduction. In *Doing cultural geography*, ed. P. Shurmer-Smith. London: Sage.

Sorkin, M. (Ed.) (1992). *Variations on a theme park: the new American city and the end of public space.* New York: Noonday Press.

Stefanou, J. (2000). The contribution of the analysis of the image of a place to the formulation of tourism policy. In Helen Briassoulis and Jan van der Straaten (Eds.), *Tourism and the environment: regional, economic, cultural and policy issues, revised second edition*, pp. 229-238. Dordrecht: Kluwer Academic Publishers.

Taylor, J. P. (2001). Authenticity and sincerity in tourism. *Annals of Tourism Research*, 28 (1), pp. 7-26.

Taylor, J. (1994). *A dream of England: landscape, photography and the tourist's imagination.* Manchester: Manchester University Press.

Terkenli, T. S. (2002). Landscapes of tourism: towards a global cultural economy of space? *Tourism Geographies*, 4 (3), pp. 227-254.

Terkenli, T. S. (2001). Towards a theory of the landscape: the Aegean landscape as a cultural image. *Landscape and Urban Planning*, 57, 197-208.

Terkenli, T. S. (1995). Home as a region. *The Geographical Review*, 85 (3), pp. 324-334.

Tress, B., Tress, G., Decamps, H. and d'Hauteserre, A.-M. (2001). Bridging human and natural sciences in landscape research. *Landscape and Urban Planning* 57, pp. 137-141.

Wang, N. (1999). Rethinking authenticity in tourism experience. *Annals of Tourism Research* 26 (2), pp. 349-370.

Weaver, A. (2000). Passenger ships as nautical vacationscapes: postmodernism, space, and multinational capitalism. Paper presented at the annual meeting of the Association of American Geographers, Pittsburgh, PA, USA, April 6.

Webster's ninth new collegiate dictionary. (1983). Springfield: Merriam-Webster Inc., Publishers.

Williams, S. (1998). *Tourism geography*. London: Routledge.

DAVID CROUCH

EMBODIMENT AND PERFORMANCE IN THE MAKING OF CONTEMPORARY CULTURAL ECONOMIES

Scene in Eleftherios Venizelos Airport, Athens, Greece, source T. S. Terkenli, 2005

T.S. Terkenli & A-M. d'Hauteserre (eds.), Landscapes of a New Cultural Economy of Space, 19-39.

INTRODUCTION

Making sense of new cultural economies, it is argued, needs consistent attention to the resonances of individual lives. Otherwise, a discussion of cultural economies remains suspended in a detached virtualism (Miller, 2000). The idea of the remaking of geographies and cultural economies remains, necessarily, a consistent search to make the subject dynamic in its resonance with the contemporary world. In recent debates concerning the reframing of the cultural economies of geography, there is an evidence of increasing acknowledgement of the overlooked importance of subjectivities within geographical explanation. This has often been difficult when trying to attend to the large scale apparent dynamics of change. The shift of geographies to focus upon cultural economies combines two profound threads that inform this chapter: the acknowledgement of the breadth and inclusivity of what economies are and the refusal mutually to isolate the cultural and the economic. Thus the economic becomes engaged and even framed in relation to the cultural, and vice versa. Such an appraisal makes more robust the limits of 'either – or' claims from these two grounding components of geographical thinking and its representation of the world. These themes are sustained in different ways across the chapters of this book.

This chapter seeks to build a critical discourse concerning space, embodied practice and lay knowledge. It does this in order to address the mechanisms through which individuals are engaged in the processes of new cultural economies. Bearing in mind that these economies are widely influenced by institutions and broad flows of complexity, it argues that space is simultaneously produced, transformed and refigured in a complex relatedness between the culturally pre-figured in commercially and institutionally mediated meanings and values, and individual and collective lay geographies. It thus contests familiar claims surrounding the continuing deconstruction of identities, human activity and meaning processes and its replacement by an abstract notion of the individual as 'consumer' in an abstract notion of virtualism (Miller, op cit). Central to a quest for explaining geographies of cultural economies, and their constitutive processes, would seem to be a breaking down of one-dimensional factoring of what makes significance, meaning, and therefrom, perhaps, identity. Moreover, these geographies are mutually circulated though contemporary complexities situated in cultures where individual practices and knowledges are conceptualised as central and institutional as other 'contexts'. Economies are conceptualised as relations through which meanings, values and identities are figured, refigured and circulated. Landscape is conceptualised as a component of this process and acts of making of space through everyday actions are mutually engaged in multiple human flows. It is argued that these flows include both ephemeral mobilities of various and diverse temporalities. Explanations of these meanings are articulated through the text's narrative.

This discussion is sourced through reflections from several empirical investigations in order to ground theoretical development; to articulate the components and flows sketched in the conceptual stage and to work critically through the possibilities and limitations of such an approach. In conclusion, the chapter offers a conceptual approach that pursues a reading of processes of cultural economy, geography and space. This conclusion does not argue for an uneven shift from one set of reasoning to another, between mediated contexts and encounters, but rather an engagement, unevenly, between reasonings. Thus, for example, it points to the importance of flows, but less in terms of being overwhelmed by our cultural contexts and their fluidity than our reticulated and dynamic engagement in them. Such an adjustment in the ways we may conceptualise the production and circulation of meanings in contemporary cultural economies suggests a re-positioning of power. Conventional understanding of 'power' is disrupted. Power emerges as a component of complex, unevenly shared processes where individuals are actively involved, not least through their making sense and affect through their life encounters (Thrift, 2004).

BACKGROUND PAPERS

This chapter will sketch a complexity of cultural economies that seeks to engage the realm of everyday life and its activities, meanings and values amongst individuals-at-large, as it were. However, first, it is important to position these in terms of key threads concerning contemporary cultural economic change. There are several core components of the prevailing arguments concerning what is becoming 'new' in contemporary cultural economies, notably concerning globalisation, complexity and mobility. The complexities tend to be understood in terms of a nexus of institutions and their circulated meanings they produce (Urry, 2003). Combined with time-space compression, this process of globalised complexity accelerates the circulation and distribution of meaning in contemporary culture, and 'real distance' diminishes, perhaps erases, the significance of particular spaces (Harvey, 1990). Cultural de-differentiation is presented as providing a weakening of the identification that individuals hold in relation to spaces and their lived experience, rendering spaces alienating from human life and emptied of meaning previously held. It has been argued that new spaces are produced that are empty of meaning and are thus non-spaces as they have not been constituted through human activity and its lent cultural meaning and value (Augé, 1995). Human action becomes marginalised in the production of meaning.

The significance of this space creation process is, arguably, experienced more widely in the outsider-ness of human life in its playful post-identity activities of gazing rather than engagement (Urry, 2002). It is also experienced in mobilities signified in the fleeting and implicitly unstable that thereby determine the further superficiality of contemporary life, and necessitously requires a ready-made reference point from elsewhere, outside the life practices of the individual (Urry, 2002). Thereby the world is engaged and consumed as meaningful through processes of detached consumption. In this process, the producers of objects of consumption and their advertising, and as an abstraction the object of consumption, increasingly produce meaning, reference, context and, thereby, operating as the key players in dynamic cultural economies. Similar complexity and time/space compression and their effects on investment and circulation affect the production sphere (Lash & Urry, 1994).

Baudrillard argued the importance of 'strategies of desire' (1981, p. 85) through which consumers' needs are mobilised, provoked, their nascent interest captured in a process of consumption before consumption. These strategies, he argued, consist of the signs on which the value of products are conveyed in the process of seduction. Baudrillard's strategies *for* consumption are crucial points in his version of cultural economies. The effective and affective power of signs are displayed and systematised through their communication. The power shifts from the objects themselves – and the subjects of their consumption – to their circulation in representations, their fuller consumption dominated by their sign-value, their value invested in anticipation. Cultural industries, the media and so on reframe life`s meanings and produce the cultural economies. New cultural economies become constructed and constituted thereby, where individuals merely 'sign up' [sic] to abstracted bundles of meaning.

The argument for a world constructed and constituted through representations, produced in the contexts of products and media, combined with contemporary social and cultural detachment, where individual, human experience, is at the margins of the contemporary world and its processes. Through various strands of these debates in uneven combination there emerges a prevailing discourse on space, place and geography. A particular English version of landscape has conceptualised landscape as text, held in representations produced elsewhere from the lives of individuals who may use or visit, characterised by perspective and surveillance by 'others' yet where the individual may also be spectator and interpreter appropriated in the gaze (Mcnaghten & Urry, 1998). Space is arguably produced and its meaning thereby constructed through media and product design, identification and promotion (Crouch & Lubbren, 2003).

Space that bears significance is presented as largely outside individual influence, power and processes. The potential for virtual 'contact' releases the need to position individuals in actual material spaces. Space, in general terms, becomes something that may offer objects of play as detachment and without responsibility. Identities are 'bought into', constructed or constituted neither in continuities of social distinction or through individuals' lives and life practices. Similarly the body in space emerges as an objectified component of this play, object of interest, curiosity, desire (Featherstone & Turner, 1995). The emerging emphasis on contemporary cultural economies tends to work with a prioritised and privileged understanding of vision as detachment, the power of representations and the alienated power and the marginal role of individuals except as support cast. The complex nexus of institutions is made more intelligible, if more complex, by a renewed attention to connectivities, where individual institutions are seen less as distinct but more inter-related in ways beyond those of protocol, production, ownership (Amin, 2004). These elements contribute to the complexity of cultural economies in reconstruction. However, there are other complexities emerging from different discourses.

In his recent discussion on virtualism, Miller has identified the incompleteness of cultural economies abstractly understood in terms of the marketplace (2000). Lee has pointed to the cracks and fractures, the incompleteness of this abstracted virtualism through which to understand a market constituted and mediated cultural world (Layshon, Lee & Williams, 2002). The debate concerning 'the local' and 'the global' in cultural terms provokes the possibility of reaction; and alternative dynamics in the making of cultural economies. Where there are other opportunities, where the individual is positioned inside some of these processes, the language used in interpretation is sometimes, ironically, borrowed from new technology and, for example the computer, the individual 'dragging' and 'indexing' significance into contexts of action (Rojek, 1997).

Whether such language enables us to get closer to understanding human processes is uncertain. To move outside these prevailing constructions and contexts may include alternative possibilities in counter-practices and resistance, as explored in the book edited by Pile and Keith (1997). Alternative cultural economies may be envisaged in terms of the organisation of life practices and exchange of objects through which different meanings may be produced and circulated (Lee, 2002; Gregson & Crewe, 1995). It may be, however, that the scope for diversity in the production and circulation of meaning may emerge through more diverse, less explicit acts of making alternative economies, or of making resistance, and it is to this arena of possibilities that this chapter now turns. In their discussion on the economies of space Lash and Urry point to the reflexivity of the consumer, the visitor, the individual engaged in the world of produced spaces (1994). The processes through which individuals may become participants in the production and circulation of meanings in their life-practices and life-exchanges are considered in the following section.

In a series of contributions, Massey, Thrift and others have argued for a closer consideration, in geography, of ways in which individuals may diversely participate in the world around them, in its economies and cultural production (Massey, 2004; Thrift, 1997; 2004). Crucially these contributions counterbalance the discussed prevailing positions of complexity, time-space compression and detachment. Moreover their emphasis is less in terms of strategies of resistance than of understanding the ways in which individuals practice their lives. In doing so especially Thrift has engaged so-called 'non-representational theory' (1997). This chapter sketches an explanation of the ways in which we may address, and make sense of, the work of individuals in making new cultural economies and of how space is engaged in this process. Thus it is argued that attention to complexity, connectivity and process needs to be 'brought down' - or up- to the encounters, actions, knowledge and ontologies of individuals in order to address the complexity and interactions through which contemporary cultural economies and their spaces may be produced.

CONTEXTS AND PROCESSES

In his exploration of what makes things matter in terms of the world of objects, Miller argues that consumption is much more than the act of purchase (1998). He argues that individuals engage products they have bought in a complex set of relationships, values, uses and there from produce meaning of products through their own life actions. He calls this process of accommodating objects into one's life 'work done'. Furthermore he points out that shopping, rather than being a key in the circulation of prefigured cultural economies, is not adequately understood as a means to acquire lifestyle meaning from elsewhere. Shopping also uses praxis necessary acts through which life can be negotiated and coped with, to get by: to get what is needed; to meet friends; to look after family members (Miller, 1997). If cultural economies are not produced only through the complexities of globalisation and its projection and circulation of meanings, then how can we get closer to identifying and to understanding other components of contemporary cultural economies?

Tim Ingold's discussions of 'dwelling' is informative (2000). He identifies a process of *dwelling* whereby encounters with objects, individuals, space and the self in the doing of everyday events may also be used to negotiate and to progress life, making adjustments and negotiating meanings dynamically (Ingold, 2000; Harrison 2000). Furthermore it is through an array of relations, objects, desires and actions that we make sense of life. Human life happens, in part, in places that are partly circumscribed in relation to what individuals do. Meaning is thus progressed through a complexity of actions, events and encounters in the spaces where action and reflection take place, rather than simply constructed, constituted or directed outwith the individual's life processes. Space may become filled with human practices, perhaps influenced by events and may prompt outside life relations, enacted. What 'makes' space meaningful may not be prescribed outside human encounters. Thus, we may argue that space is 'made sense' through the ways in which it is encountered.

To consider more closely the ways in which individuals encounter space, it is informative to consider recent work surrounding practice and performance, components of what has become known, perhaps in over-simplification, as non-representational social science and humanities, including geography. The possibility of this label being over-simplified derives from the apparent projected meaning of being a process outside representation. It is not intended here to argue that actions and representations are not mutually bounded. Rather, that these are mutually inflected and, if at all, held by very porous boundaries.

Recent interest in developing ideas of human action and meaning has developed through renewed and revised attention, to the work of Merleau-Ponty (1959). He argued that thinking was part of the active engagement and relationship that humans make with and in their world, rather than some detached disembodied mental product away from the concrete world of actions and encounters that individuals make. His influential discourse on practice developed also through the work of Butler and others such as Grosz, on performance, have pointed towards the complexities through which the individual acts and engages the world, and may use these encounters to negotiate their understanding and meanings. The ways in which individuals engage the world thereby does not seem to be privileged in vision. Important conceptual discussions of reflexivity tend to confine themselves to the abstracted mental processes that may be involved (Lash and Urry, 1994). Just as Lash and Urry emphasise the importance of the reflexive self in the ways in which the surrounding world may be interpreted by individuals, so considerations of embodied practice direct attention to means through which this reflexivity may happen (1994).

The individual is multiply sensate, and so encounters numerous sources and genres of evidence in the world. Through this multiple sensuality, complexities of action engage multiple evidence – but not merely as physical interaction causing a sensory result. Sight combines, or is combined, with hearing, touch, taste and so on. Actions may occur in the numerous, often mundane processes of keeping going, of holding on to life and what it means. In a sense, this concerns getting on with life, but individuals engage also in exploration and play, escape and confrontation. The value and meaning of objects, things, relations, others and actions are infused with emotion and available knowledge and sense of the self. The individual 'reading' of surroundings is not inert, but active, dynamic, 'at work. Practice is, then, expressive, engaging or avoiding, making affect on what is done; and on the objects, including space, through which it is done.

Moreover, practice is inter-subjective, in presence or absence, as things happen 'in relation to' others, in presence or absence, familiar and unknown. Being with others, for example, can give different character to a similar event, in a similar space, lived alone (Crossley, 1995). There may also be a poetic component in the ways in which actions and encounters are made, engaging imaginatively and playfully (de Certeau, 1984; Birkeland, 1999). These components have been discussed elsewhere (Crouch, 1999; 2001). Here, it is necessary to develop these apparently mundane, apparently isolated, events into their possible value in making sense of what is happening in the dynamics of contemporary cultural economies. The human actions, events, and so on that these insights concerning embodied practice unpack are complex. And what individuals do is felt, experienced, and adjusts their engagement with the world and the way they make sense of it. Thus active practices are felt in the *doing* (Harre, 1993). Doing, the individual may figure, reframe and adjust the world. Things and spaces do not bear objective but subjective meanings, meanings made through this dynamic process of encounter in practice. That dynamic practice involves the effects and influences of surrounding, mediated contexts, inflected by meanings of things, relations and spaces embodied in representations, objects and actions. Individuals may then adjust the world in the way they perform their lives; and the significance of spaces where that life takes place.

The power and contextualisation of individuals' lives through the mediated complexity of global cultural economies, perhaps in such a state of temporal flux that the individual's power over them is further erased, may deliver the protocols that everyday existence requires. Yet in their work done, in their enaction of life, individuals may have other possibilities. These ideas can be mobilised towards thinking of what individuals do, in performance, as negotiating life, but can they thereby influence the character and shape of cultural economies? Recent debates on performance suggest further complexities in the mundane.

Performance can emphasise the framework of everyday protocols in the ways in which cultural economies are produced and events, places and actions are 'culturally' defined, prefigured, prescribed, as outward influences, and through which protocols the world is made sense of. However, recently another feature of performance suggests something rather different. Elizabeth Grosz argues that in performance there is the possibility of openness, through which life may be modulated and the self reconfigured, that life '(duration, memory, consciousness) brings (something) to the world: the new, the movement of actualisation of the virtual, expansiveness, opening up': enabling the unexpected (Grosz, 1999, p. 5), my second parenthesis). Thereby new strategies can be constituted in cultural acts, refigure the world, transformatively. Discussing the potential of the individual to affect the world and its meaning, developing Merleau-Ponty, Radley argues that individuals create a potential space in which they can evolve imaginary powers of feeling (1995, p. 14). This is like Grosz's notion of 'becoming', a competence to figure the world and shape its meaning through the complexity of human actions, akin to Ingold's notion of doing as 'dwelling' (2000, op cit). We may distinguish between ideas for things, space and so on, as prefigured and determinate, and the motor of 'dwelling' that sustains the present and future, from which contemplation and new possibilities of re-configuring the world, in flows, can occur.

Rather than consume a mediated semiotics through which they make sense of life, individuals may progress their own significations, and apply the embodied character of performance, as an *embodied semiotics* (Crouch, 2001). The category of performativity is taken further as ongoing and multiple inter-relations of things, space and time in a process of becoming, in engaging the new that may be, like Radley's consideration of embodied practice, unexpected and unconsidered, not pre-figured, suggestive of a similar performative shift beyond mundane, routine habituality. That 'going further' may emerge from exactly those apparently momentary, mundane things that we do (Dewesbury, 2000). Moreover, the borders between 'being' as a state reached and 'becoming' are indistinct and constantly in flow (Grosz, 1999), although they may be focused in *the event* (Dewesbury, 2000, pp. 487-489).

In the present discussion, 'becoming' is distinguished from 'being in' the sense of Grosz`s *becoming* as 'unexpected', where performativities may open up new, reconstitutive possibilities, beyond protocols and habituation. Protocol and becoming point to new complexities in the construction and constitution of meanings in the contemporary world. An important component of performance as performativity emerges in the ways in which individuals make and use their encounters to negotiate their lives. There is a potential fluidity between being and becoming; of 'holding on' and of 'going further'. A realisation of a state of being in security can be found in repeated performance. However, there is also the possibility of reaching forward, of going further, of exploring, trying something new, rearranging what things mean in sensation and desire. Individuals operate routinely along these tensions (Crouch, 2003). Performances in 'going further' and in 'holding on' do not, of course describe a simple polarity, but operate in multiple forms of complexity themselves.

The geographies of the world in which individuals perform are significant here. Rather than take cues from a nexus of global cultural processes, individuals may be considered to make space in a process of spacing. Practice and performance take place in a variety of different spaces, multiple spaces. Contemporary mobility and time/space compression has enlarged an available number and diversity of material and metaphorical spaces. Yet individual actions, life events and relations still familiarly happen in concrete, material spaces. Those spaces are engaged bodily, if temporally. Meaning, perhaps multiple complex meanings of space are open to fluidity amongst complex human activities. Spaces and their fragments, metaphorically and materially, can become affected, coloured, (re)figured through the expressive encounters the individual makes in relation to them.

These diverse life-spaces are constituted through performance and practice, and include the metaphorical and the concrete. The actions happening in and between spaces contribute to the ways in which the individual progresses a practical ontology (Shotter, 1993) and its constituent lay geography (Crouch, 2001). Spaces in this geography may be relatively discrete, or relational, mutually inflected. However, contemporary mobilities are often of very limited distances and time/distances. In the practical ontologies discussed individual spaces of encounter and action are significant, or may be signified, through what individuals do, how they act. In these ways embodied practice and performance engage in producing cultural economies of space. Space is not unique in the agency of performance and space, relations, events and artefacts become connectively adjusted through performance. Indeed, this discussion does not privilege the individual, stripping the influences, framing and contextualisation of wider events and meaning production of their power. Rather than delete, or even reduce, influences, the discussion acknowledges them but argues for a more active, negotiative, even combative way in which individuals engage the world and its meanings, its cultural economies (Nash, 2000).

Performativities can distinctively colour geographical knowledge. '(T)he ability to reflect consciously on thought and sensation, which are initially spatially located, comes through this symbolic dimension. This dimension is blended with space and time, for symbols are used for a means of communicating with others... a means of communicating with ourselves.' (Burkitt, 1999). Lay geographies become part of contemporary negotiations of identities, a crucial component of the progress of contemporary cultural economies. This component of identity and the diverse facets of performance in life negotiations is taken further through a consideration of five diverse narratives.

CULTURAL ECONOMIES IN PROCESS

This section uses largely empirically-informed work by this author (Crouch and Toogood, 1999; Crouch, 2001, 2003). The narratives briefly sketched are diverse, and are intended to connect different components and arenas in which contemporary cultural economies may be mobilised. The first builds a brief narrative through the work of an artist who was a member of the International Modernist Movement, working in Cornwall, in the southwest of England, during the middle of the twentieth century; the second is drawn through investigations of a mundane 'pastime' [sic] known in the UK as allotment holding, in the US as community gardening. The third narrative is worked through a component of contemporary mobility combined with tourism, caravanning. Finally, two reflective narratives sketched from the author's developing work, of the framing – and practice – of cultural economies of the Mediterranean and of American landscapes for visitors, in the UK.

The artist Peter Lanyon brought together an anarchist concern for the human condition and for social progress, alongside a critical reappraisal of artistic traditions of perspective that he sought to disrupt through his self awareness of his own embodied encounters with space. In this, he was directly influenced by his time with the constructionist Naum Gabo and abstract expressionist influences of Rothko. He felt that the complexity of the encounter he had with space[s] was detonative of experiencing a departure from everyday experience, yet his work familiarly engaged elements from everyday, mundane practice. He felt uneasy with ordered, 'perspectivised' narratives of the way in which the world and its influences are often understood, and shifted from the familiar notion of the painterly 'gaze'. In his own intimate encounters with space *around* him he found dynamic instability resonant of his sense of the world:

'The beginning of a painting may be down a mine or on top of a bus. ... an abnormal sense of rightness in the presence of something happening or place... in West Cornwall this whole existence of surfacing deep and ancient experience is obvious.'

He connects experiences:

'After a north storm ... seamen can be seen plodding the beaches and picking objects out of the sand... a fascination which has affected me. These are reassurances of the living I know in my paintings - the comparisons, the closeness and the edges of lives different in appearance but fundamental in their history .... so that the farmer, the miner, the seaman all in their own journey make outward the under things' (Lanyon, p. 292)

Rather than seek to dominate space [landscape] through perspective and order, he sought a fractured patina of encounter signified in mobility, where lines represent body turns and the unexpected rather than boundaries and limits. He felt that, however intimate, these bodily encounters were not parochial:

'Environment', or 'nature' for him was something that he constructed through his encounter with space. He engaged the body and landscape in a multi-sensuous way. He was engaged in making representations, but these would seem to be attached to, rather than detached from, his bodily, or embodied, experiences. He raised the experiences of the immediate and close that he felt to levels of significance of being in the world:

'I found an ease ... that things were happening, ... a house was standing gaunt beside me, the road as I went back into town was hedged on either side, but the sea was on one side, and... all the small grasses were moving... with a curious blowing. Now that's a sign to me that there is a fusion, ... an interest being created that connects the things growing in yourself. I can pin it to the place.. to establish itself in time and space.' (Lanyon, op cit)

The power of his feeling – and interpretation – was of becoming: dynamic, uncertain, in flow, and concealing a complexity of enquiry, of going further, engaged too in being troubled over surviving industries and human lives. He engaged, in an awkward way, wider cultural narratives with his own embodied experience.

Another entwined encounter emerges in the ways in which individuals narrate their encounters with life and life-spaces in working a small piece of land as an allotment 'garden'. These individuals are not engaged in making explicit representations, but engaged in embodied practice, through which they construct their sense of the world and its relations. The allotment has, in the UK as elsewhere, a significant but diverse political history from historic struggles over the right to land; efforts to make distinctive identities amongst the poor; to balancing complexities of contemporary life and work, for example in working in new technology industries; seeking ecological engagement with space (Crouch & Ward, 2003). Yet, for most individuals, what they do is not an overt political act, but to pass time and do something they enjoy, as these notes show.

> 'My allotment has enabled me to find a side of myself I did not know existed and it also helps me to cope with an extremely stressful job in a stressful city.' (Yvonne, 35, Weardale, 1997).

Talking over their experience of working adjacent plots on an allotment site one gardener replies of their interaction:

> 'And I've learnt from him. I've learnt some ways of planting, I've learnt real skills about planting, Jamaican ways of growing and cooking. And I've also learnt about patience and goodness and religion, too. It all links in. ' (Alen, 65, Birmingham, 1994).

Another plotholder identifies what doing this space means to him:

> 'If you give somebody anything they say – where did you get it from – I say I grew it myself. You feel proud in yourself that you *grows* it, you know, if you get it from the shop, some of it don't have any taste. On the allotment you plant something and it takes nine months to come to' (Len, 70, quoted in Crouch and Ward, 2003).

To a degree, it may be argued that there is a distinct 'alternative cultural economy' at work here outside human encounters. Yet, their making sense of life is articulated in more complex ways: in terms of individuals encounters, engagements and values that, for some, are characterised by sharing materials and work, co-organising, sometimes defending sites from commercial development, and spending time together. Can something similar be at work in a less culturally distinct encounter, through individuals who make leisure, and tourism, by going in a caravan, near home, or overseas, who use sites on a commercial basis?

Yet, do their values and meanings that surround this space and action comply with a mediated cultural economy of globalisation? Caravanning can be highly commercialised and many trade, leisure and club magazines devoted to the activity promote equipment and commercial sites to be visited (Crouch, 2001). In part, these commercial mediations pre-figure a dream of freedom, of being able to visit places otherwise inaccessible and, where individuals tow their 'van or drive mobile homes, to experience a free-wheeling choice of locations.

This leisure and tourism activity is often characterised through its mobility, yet mobility can bear diverse meanings. To an extent, mobility is to arrive, at a space of distinct cultural and bodily encounter, where the individual feels enabled to negotiate his/her life differently. One caravanner, Tim, describes the importance of what space can offer:

> 'When you pass C..., life feels different.... Caravanning, it all makes me smile inside. I mean, everyone just comes down to the ford and just stands there and watches life go by. It's amazing how you can have pleasure from something like that. I just sit down and look and I get so much enjoyment out of sitting and looking and doing nothing. We wake up in the morning, open the bedroom door and you're like breathing air into your living'. (Tim, 45, Weardale, 1997)

He discovers feeling by doing. His performance comprises haptic vision, caring, relating and finding it uplifting; touch and other performativities as he feels his body encounter the space between the van and the water's edge that is suspended as he revels; he smells as he breathes what feels like different air. He is aware of these and constructs his representation of what he does, through an interpretation of how he constitutes space as spacing. The space becomes his own, through what he does and the way he does it. He gathers his sense of what he is doing, making things significant, through these components of performance, not only from other-figured information. In his narrative, he communicates a making sense of the world through what he is doing (Crouch, 2003).

It would be easy to over-emphasise the significance of these actions and their narratives, even though those involved either habitually participate or make journeys that they regard as very significant in their lives. Yet, similarly, it may be an exaggeration to envisage these self narratives of what seemingly 'matters' to individuals as being merely trivial. The next section explores the possibilities of thinking through these components of individual encounters, in relation to discourses concerning the making of contemporary cultural economies.

Turning further to what may be regarded as powerful components of contemporary global economies, more 'obviously' mediated cultural economies and the spaces that may convey them, the following paragraphs explore reflections on the complexities of place-construction and representation, and the value and meaning given to practiced, performed, spaces. Two kinds of geography are considered, briefly, here. One is the cultural economy through which the Mediterranean is constructed and constituted, and thus prospectively consumed. The other is the cultural economy through which 'American landscapes' are produced for consumption in Britain (Selwyn,1996; Campbell, 2004).

The Mediterranean has been significant in diverse ways in British culture for several centuries (Inglis, 2000). In terms of its artistic and commodified representations, the Mediterranean is a bundle of significances for contemporary Britain. These significances have different temporalities. The Mediterranean may be consumed through the pre-figured circulation of an iconography of culture and religion, each of which has deep, long-term histories. More recently, especially in the artwork of the early decades of the twentieth century, these often contested histories have been combined with new layers of the exotic, sexual and sensual temptation and risk borne on depictions of harems and their bodily display as interpreted by artists (Lubbren, 2003). Since the mid twentieth century, newer versions of heat, risk, the body and the exotic have been inserted in more popular discourses as well as appeals to diverse kinds of tourists to engage in delights of their bodies on beaches and elsewhere, very recently emerging in versions of clubbing life.

Destinations of tourists such as the Mediterranean may be given meaning through performance as well as through their prior significations too. In her autobiographical accounts of both performing Queensland's Bondi beach and the uplands of the British Pennines, Ann Game discusses her bodily encounters as reaching things that cannot be seen and making what she calls *material semiotics* 'in the doing' (1991). Through reflection on the imagery that promotes Bondi, Game identifies in her own experience something very different. For her, it is the intimate engagement with the features of the beach, the elements, a book to read, her own memories of doing similar things on this beach years ago, that constitute her encounter. Rather than its ostensible features pre-constructing the contours of experience, body-practice can be significant in the way space is constructed and constituted, through bodily, imaginative encounters that may also make playful use of memory and time (Crang, 2000).

In a different way, but still suggestive of this interplay, Andrews' investigation of charter tourists in Magaulf and Portonova, Majorca, unpicks individual and group efforts to reconstruct an idea of Britishness- racist, sexist, militarily patriotic, alienating. Whilst the destination she examines is known for components of Britishness, including English breakfasts and enclaves of British, especially English, ex-patriots, it was significantly in the doing, the embodied encounters in clubs, on the beach, in bars, that the idea of Britishness is mobilised. Mobilisation happens in body-actions, both as display, but also in the physicality of often very expressive, heavily inter-subjective and 'social' encounter (Andrews, 2004). The individual tourist is re-interpreted as performing, enacting space-ing, her performance produces an embodied semiotics, in an ongoing and fluid time-space encounter. As Game shows, the projected imagery does not hold intact. The individual may make the trip, encounters space and may make his/her own sense of space in his/her own, hybrid, compilation. This would suggest that practical ontology can refigure prefigured cultural economies of these places, and the values and meanings through which relations are constituted through visits such as these may be at variance with what are often considered to be prevailing meanings and their significations.

It is familiar to assert the power of distinctive commercialised identities in the mediation of America through holidays and the designs of destinations and the signification of their globally available products as exemplary of processes through which contemporary cultural economies of America, and more generally of the West, are produced and circulated. In a way similar to the commercialised cultural projections of The Med, this kind of explanation works in a linear fashion to produce an experience of consumption, experience and place that is projected through global commodification and politically constructed meanings and values. The American Dream, western ideals, freedom and superiority, global commercial and lifestyle power, are arguably signified in the landscapes that are intended to contain these representations in a simulacrum of power, product and positioning (Campbell, 2004). Yet how do these supposed projections 'work' in the experience, in the embodied encounter, that individuals may make in the spaces that are enveloped, in design terms, with American stylisation? Campbell speculates on the limits of how we understand the cultural circulations of America thinking through the encounter the individual makes. Jewesbury interrogates a diversity of meanings and identities engaged in a visit to London Bridge in Arizona (2003). It is difficult to be secure in an argument that presumes linear follow through of the construction of such a cultural economy as 'America'.

Projected identities of The Mediterranean and of America only go so far in making offerings of cultural identity and identification; the proscribed and pre-figured, it is argued, may only incompletely be grasped through a discourse that regards the consumer, the tourist, the individual, as 'in receipt of', rather than as part of, the processes of contemporary cultural economy components. The power of commercially and institutionally mediated cultural economies operates in terms of influences upon (rather than determining or framing) cultural economies. Embedded in their processes are the important components of how knowledge is made as discourse and the circulation of that knowledge in and through what individuals do, feel and think. In processing an individual, and subjective, practical ontology and its concomitant reflexivity, lay, or everyday geographical knowledge happens through which the individual may be able to work his/her own signification, to work in ways different from a projected cultural economy 'from without'.

Through these varied considerations: artwork; using a plot of ground; touring; visiting places enveloped in powerful imagery, the work of the individual, in embodied encounters, is dynamically involved in cultural economy processes. Contextual, mediated signification is evident in each, but is apparently often used as a tool amongst the actions and events in spaces encountered in what individuals do, in their dwelling. Thus, individuals make connections and dissonance and negotiate, make sense of things, and progress their lives. Through performance, they can engage, discover, open, habitually perform and enact, reassure, become, create in and through the performance. In things done there is frequently a tension between these components of holding on and going further, between reassurance and the excitement of experiment. Individuals build their agenda through what they do, using available resources. This is the process of their practical ontology as cultural significance and meaning are negotiated and circulated amongst others inter-subjectively.

In the following section, these observations are worked through a series of considerations that relate back to prevailing directions in the discussion of the dynamics of contemporary cultural economies. These discussions include complexity and the so-called 'global'; time/space issues; identity, networks of cultural economy production, and flows; the signs through which cultures are developed and communicated. Space is articulated and problematised in these different components. A key emerging issue appears to be the articulation of complexities between institutional/commercial frameworks of cultural economies and the complexity of dwelling that may be regarded too as a framework of a dynamic cultural economy sketched in this chapter.

CONNECTIONS AND SPACES

In the so-called 'aftermath' of socially transmitted identification it is often argued that commodification and linear processes of consumption-production provide major conduits through which contemporary meanings and identities are forged, adopted, made. Through an attention to bodily action and knowledge, Burkitt positions identity in much more subtle and complex arenas, of individual encounter and practical ontology, and in the relationships between these and 'contexts':

'Indeed, I would argue that this is how most people function in their everyday lives: by acting on the basis of a sense of what ought to be done, drawn from experience of previous situations and the tacit knowledge this has developed, along with 'gut feelings' about which is right in the circumstances. Most people refer to this as intuition or common sense, where it is actually a complex interplay of knowledge and feeling' (1999, p. 149).

He situates this in terms of the relationality between embodied lived encounters and mediated contexts, and time/space:

> '... because the symbolic is another dimension of time and space it can never be completely separated from [the sensual]. ... Could we understand an abstract idea without a bodily sense of experience by which to place it and give it meaning, even if the idea ran contrary to embodied experience?' (Ibid, p. 150)

Events, actions, encounters, still happen in concrete spaces. It may be unhelpful to continue to talk in terms of time-space compression. It may be argued that significant occurring processes may be better termed time/space complexity or uneven time/space connectivity. Time-space complexities would seem to capture the possibilities and limits on offer and seem to avoid over-simplification. There are multiple connections across multiplicities of encounters in flows in multiple spaces that operate unevenly. The lives of individuals may be multiply-situated – or not – and yet spaces of everyday performance or practice may be given significance. In another way, global knowledge/travel mobilities may have very uneven affects on ways in which the world is made sense of and identities progressed (Clifford, 1997; Crouch, Aronnson & Wahlstroem, 2001), across different flows of memory and time, often crucially 'in negotiation' rather than linearly given emphasis (Crang, 2000). For example, spaces part-constructed through colonialism may be also enacted and constituted, refigured, through the experience, feelings, expressivity of individuals. Advancing interpretations of cultural economy dynamics deserves close interrogation of the mutual flows amongst all these plural spaces.

Whilst much is made of mobility that is global, much human mobility occurs in short distance and intimate encounters. Of course, both need to be engaged in any contemporary analysis of space and its cultural economies. Wider networks of enacted spaces and their representations are often 'background' rather than focal. The power of, for example, global mobility as travel, working practice, even migration, even if fairly habitual, may be exaggerated in terms of its power to refigure lay geographical knowledge and identities. The boundaries between interim spaces of 'origin' and 'destination' may not be vapourised, and making sense of life between its multiple spaces may be more one of negotiation and complexity than replacement. Sites and spaces of the virtual – in terms of enacted chatrooms and games - are encountered in a room, at a desk, interrupted by coffee and use of other facilities, even human encounters, in flows of multiple life encounters of varying intimacy and distance.

Complexity in human space-ing encounters may provide potentially useful modes of explanation. 'New' spaces may be more relative, or rather related, relational in the lives and encounters, identities and exercises of power in terms of knowledge and its enaction than absolute. Investigations and critique of examples of cultural economies would benefit from engaging this complexity of human action, encounter, lay geography and practical ontology. Rather than time and space compressed, it is time and space ruptured, multiplied, temporally unravelled, engaged and negotiated at a human scale.

It is inadequate to consider institutionalised spaces to be dominating frameworks through which cultures and knowledges, attitudes, meanings and identities are sustained and circulated, as there are numerous other spaces in which individuals act and knowledges in which they are engaged. Commercialised spaces may not be externally 'given', as these are filled with meaning constituted through individual encounters. Whilst in a fascinating discourse on non-places, Augé discusses so-called 'empty space' of such as airport lounges, which they can be, these may be filled with emotion, despair, loss, excitement and anticipation of arrivals, love, anxiety. These spaces are difficult to understand, in terms of human encounters, as adequately 'empty'. The world, its landscapes/spaces may be trivial, their practices mundane, yet such a claim remains critically to be checked in life circumstances. The contextualised disposability of space and consumption practices may be of less significance significant. This suggests variability, diversity and complexity.

It is often argued that one component of the contemporary cultural economy is influenced by a globally-constructed understanding of environmental concerns, in terms of lost diversity and damage, that emerges in local action. Yet individuals come to lay knowledge of what environment or nature mean through direct experience too (Crouch, 2003a). Reading, seeing and knowing what happens across the world is engaged in intimate encounters, connected. Rather than ' think globally-act locally ' the thinking and action is more complex (Burkitt, op cit). Through their practical ontologies individuals have capacities to affect the world, to act, to combine, to influence, through both negotiating the world to make sense of it through their lives and through reappraisal. They affect change, not merely through resistance and alternative practice, but through the progression of identities, everyday relations and acts (Crouch, 2003; Thrift, 2004).

CONCLUSION

The relations, production and circulation of meanings appear often to be interpreted as being driven by complex institutional processes. However, there emerges a case to this argument with a discourse of more dispersed, differently complex, processes of cultural economies. Through a consideration of recent theoretical engagements between so called representational and non-representational geographies, this chapter has developed from a discourse on the processes of individuals' encounters with space and their participation in making mediated cultural economies to ideas of practical ontology, identity and power; power in building and making sense of the world. A discussion on complexities of developing cultural economies has been sketched that seeks to acknowledge the non-financial and non-institutional components in the production, distribution and consumption of meanings, values and relationships that surround the complexity of diverse components of human actions and subjectivities through which the world is both made sense of and progressed. In their wide-ranging discussion of economies of signs and space, Lash and Urry discuss possible ways in which 'the aesthetic reflexivity of subjects in the consumption of travel and the objects of cultural industries create a vast *real economy*', through their engagement in the flows of diverse cultural producers and mediators (1994, p. 59, my emphasis).

Yet, this 'real economy'- of art galleries, bars, taxi drivers and brokers – conceals the complexity of cultural economy. The different components, spaces and events used are combined through the performative encounters and practices individuals, temporarily consumer or traveller, makes. Through their performances, individuals work, selects significance amongst a complexity of things, feelings, relations and actions, in a process of affect. The result may be a cacophony or patina, negotiated, different spaces given embodied significance. The apparent 'real economy' is worked through and mediated by individuals into their own relation with these other resources of economy. Prevailing discussions of cultural economies and complexities, as well as more recently, of connectivities, has tended to underplay these more human components of contemporary cultural and geographical worlds.

It is important that a perspective on the production, circulation and figuring of cultures reattaches to the individual, rather than positions the individual detached from the production, creation, construction and constitution of the world and its components and meanings. As such, this chapter has engaged the institutional, and the reflexive individual, components of the cultural economies of time and space by engaging the cultural dynamics of contemporary life not 'inside' but rather 'in relation to' institutional and other broad flows of global processes and pressures. This discussion considers not just the important area of intentionally 'alternative economies acting outside, or in resistance to, prevailing economies. It argues the value of conceptualising landscapes of new cultural economies as produced and progressed through the work of individuals in a way semi-attached to mediated cultural contexts. Thus, it becomes possible and necessary, to disrupt conventional framings of 'the economic' and the institutional construction of the cultural further, in terms of space. At the same time, it is necessary to be cautious concerning the notion of 'the new'. The tracking of what may be 'new' may be less a response to changes in the world than new responses in the academic world to making sense of processes, power and change.

Landscape emerges from this discussion as a relational and unstable component of this encounter process, of an ongoing spacing of the world, and as ever-temporal, material and metaphorical existence. Space is ever-temporal and also multi-temporal, in terms of the multiple and often not complementary flows of time and memory in the production of making sense (Crang 2000). The significance rendered to landscape here works from its destabilised role and as process, in making sense of, and negotiating, the world in the figuring and refiguring of cultural economies.

Acknowledgements: Versions of this paper were presented at the IGU-IBG conference, Glasgow 2004, and at the Conference: Locating Cultural Economies, Kalmar, Sweden 2004.

Tourism School, Derby University, U.K.

38 D. CROUCH

REFERENCES

Amin, A. (2004). Regions unbound: towards a new politics of place. *Geografiska Annaler* B.86.1, pp. 33-45

Andrews, H. (2004). *Escape to Britain: the case of charter tourists to Mallorca.* Doctoral Thesis, London Metropolitan University unpublished

Augé, M. (1995). *Non-Places; an introduction to a theory of super-modernity.* London: Verso.

Baudrillard, J. (1981). *For a Critique of the Economy of the Sign.* St Louis: Telo.

Birkeland, I. (1999). The mytho-poetic in northern travel. In D. Crouch (Ed.), *Leisure/tourism geographies* (pp. 17-33). London: Routledge.

Burkitt, I. (1999). *Bodies of thought: embodiment, identity and modernity.* London: Sage.

Butler, J. (1997). *Excitable speech: a politics of performance.* London: Routledge.

Campbell, N. (2004). Producing America: re-defining global tourism in a post-media age, in Crouch D., Jackson R., and Thompson F. (Eds.), *Convergent cultures: the media and tourist imaginations* (pp. 198-214). London: Routledge *forthcoming*

Clifford, J. (1997*). Routes: travel and translation in the late twentieth century.* Cambridge, Mass: Harvard University Press.

Crang, M. (2001). Rythms of the city: temporalised space and motion. In J. May and N. Thrift (Eds.), *Time/Space: geographies of temporality* (pp. 187-207). London: Routledge.

Crossley, N. (1995). Merleau-Ponty, the elusory body and carnal sociology. *Body and Society* 1,43-61.

Crouch, D. (1999). The intimacy and expansion of space. In D. Crouch (Ed.), *leisure/tourism geographies* (pp. 257-276). London: Routledge.

Crouch, D. (2001). Spatialities and the feeling of doing *.Social and Cultural Geography* 2 (1), 61-75 .

Crouch, D. (2003). Spacing, performing and becoming: tangles in the mundane. *Environment and Planning A* 35:1945-1960.

Crouch, D. (2003a). The performance of geographical knowledges. In Szersinski B et al (Eds.), *Nature performed.* Oxford: Blackwell. pp. 17-30

Crouch, D., Aronsson L. and Wahlstroem L. (2001). The tourist encounter. *Tourist Studies* 1, 2. pp. 253-270

Crouch, D. and Lubbren N. (2003). *Visual culture and tourism.* Oxford: Berg.

Crouch, D and M. Toogood (1999). Everyday abstraction in the art of Peter Lanyon. *Ecumene* 6 (1), pp. 73-89; 196-213.

Crouch, D. and Ward C. (2003). *The allotment: its landscape and culture.* Nottingham: Five Leaves [4th edition].

De Certeau, M. (1984). *The practice of everyday life.* Berkeley, Ca.: University of California Press.

Dewesbury, J-D. (2000). Performativity and the event. *Environment and Planning D: Society and Space* 18, pp. 473-496

Game A. (1991). *Undoing sociology: towards a deconstructive sociology.* Buckingham: Open University Press.

Gregson, N..and Crewe, L.(1997). The bargain, the knowledge and the spectacle: making sense of consumption in the space of the car-boot sale. *Environment and Planning D: Society and Space* 15, pp. 87-112

Grosz, E. (1999). Thinking the new: of futures yet unthought In E.Grosz (Ed.), *Becomings: explorations in time, memory and futures.* Ithaca, NY: Cornell University Press 15-28 .

Featherstone, M. and Turner ,B.S.(1995). Body and Society: an Introduction *Body and Society* 1,1:1-12

Harre, R. (1993). *The discursive mind.* Cambridge: Polity Books.

Harrison, P. (2000). Making sense: embodiment and the sensibilities of the everyday. *Environment and Planning D: Society and Space* 18, pp. 497-517.

Harvey, D. (1990). *The condition of postmodernity: an enquiry intto the origins of cultural change.* Oxford: Blackwell.

Inglis, F. (2000). *The delicious history of the holiday.* London: Routledge.

Ingold, T. (2000). *The perception of the environment: essays in livelihood, dwelling and skill.* London: Routledge.

Jewesbury D. (2003). London Bridge in Arizona. In D. Crouch and N. Lubbren (Eds.), *Visual culture and tourism* (pp.223-240). Oxford: Berg.

Lanyon, P. (1962). In Lanyon, A. *Peter Lanyon* 1993, Newlyn Cornwall: Lanyon private publication.

Lash, S. and Urry, J. (1994). *Economies of signs and space.* London: Sage.

Layshon, A., Lee, R. and Williams, C. (2003). *Alternative economic spaces.* London: Sage.

Lubbren, N. (2001) *Rural artists' colonies in Europe: 1870-1910.* Manchester: Manchester University Press.

Mcnaghten, P. and Urry, J. (1998). *Contested Natures* London: Routledge.

Massey, D. (2004). Geographies of responsibility. *Geografiska Annaler* B: 86,1,5-19.

Merleau-Ponty, M. (1962) *The phenomenology of perception.* London: Routledge.

Miller, D. (2000) Virtualism: the culture of political economy. In I.Cook, D. Crouch , S. Naylor, & J.Ryan (Eds.) *Cultural turns, geographical turns.* London: Longman.

Nash, C. (2000). Performativity in practice: some recent work in cultural geography. *Progress in Human Geography*, 24, 4, pp. 653-664

Pile, S. and Keith, M. (Eds.) (1997). *Geographies of resistance.* London: Routledge.

Radley, A. (1995). the elusory body and social constructionism. *Theory, Body and Society* 1 (2), pp. 3-23.

Rojek, C. (1997). Indexing, dragging and the social construction of tourist sights In C. Rojek and J. Urry (Eds.), *Touring cultures: transformations of travel and theory.* London: Routledge.

Selwyn, T. (1996). *The tourist image.* Chichester: Wiley.

Shotter, J. (1993). *The politics of everyday life.* Cambridge: Polity Press.

Thrift N. (1997). The Still Point: resistance, expressive embodiment and dance. In S. Pile and M. Keith (Eds.), *Geographies of resistance* (pp. 124-154). London: Routledge.

Thrift N. (2004) Intensities of feeling: towards a spatial politics of affect. *Geografiska Annaler B:* 86, 1, pp. 57-78.

Urry, j. (2002). *The tourist gaze.* London: Sage.

Urry, J. (2000). *Sociology beyond societies: mobilities for the twenty-first century.* London: Sage.

Urry, J. (2003). *Global complexity: an introduction.* London: Sage.

BASTIAN LANGE

LANDSCAPES OF SCENES

SOCIO-SPATIAL STRATEGIES OF CULTUREPRENEURS IN BERLIN

Greige office temporally reprogrammed, featuring video artists and electronic music in their self – designed "sushi-bar", source B. Lange.

Abstract: One of the key urban and cultural developments in post-reunification Berlin is the emergence of new hybrid cultural as well as entrepreneurial agents (the so-called *Culturepreneurs*, see Davies/ Ford 1999). Germany's new capital has been suffering under continuous socio-economic crises for the last 6-8 years. In this situation, the catchword of a «new entrepreneurship» alludes to individualised marketing strategies and social hardships, but also to skilful alternation between unemployment benefit, temporary jobs, and self-employment structures as practiced by numerous young agents in the field of cultural production.
Semantically supported since 1998 by terms like «Generation Berlin» (Bude, 1999), new entrepreneurial trajectories and professional careers in the field of popular culture, the production

41

T.S. Terkenli & A-M. d'Hauteserre (eds.), Landscapes of a New Cultural Economy of Space, 41-67.
© 2006 *Springer. Printed in the Netherlands.*

of symbolic goods, and the new digital media-oriented economy of Berlin, can be analysed from an individual perspective. In the last years, declining financial support from the government forced many cultural activists and artists to open their professional practices towards corporate firms, new forms of project-based cooperation as well as specific spatial practices in order to survive economically, culturally and socially remain on the targeted market.

Furthermore, *Culturepreneurs* practice a playful attitude using distinct locations and specific places to gather temporally different scenes. Places are the terrain of the post-industrial city where different and heterogeneous professional scenes engage to provide various contexts ensuring processes of socio-spatial re-embedding. Cultural codes, urban myths and local narratives, often invented by *Culturepreneurs*, are subtly mixed at distinct places. Thus, social ex- and inclusion and thereby the uniqueness of their popular cultural and artistic products is ensured. The ability to «read» and «experience» these urban styles and event-oriented conditions (clubs, music, galleries etc.) is a prerequisite for temporary scene participation in heterogeneous urban landscapes. *Culturepreneurs* play a decisive role, acting as key agents by providing and inventing new urban narrations.

The interpretation of these new spatial strategies – based on qualitative empirical research - offers insights in new models of urbanity shaped by design-intensive professional group members and their associated locally based scenes in the fragmented and moving city of Berlin.

BERLIN AND ITS NEW CULTURAL ENTREPRENEURIAL GENERATION

Defining the novel: concepts and perspectives

In the wake of the political formation of a «Neue Mitte» in Germany, a «New Centre», in 1998, it is now, by a process of diagnosis, possible to establish connections between political strategies and socio-cultural urban development processes. With the paradigmatic political change of 1998 in Germany, individualistic, entrepreneurial qualities, which interact in a new way with the urban, came to be expressed. One potential champion of these developments is a structural type, which has yet to be defined in greater detail, the *Culturepreneur*.

First of all, the term *Culturepreneur* is a compound of *cultural* and *entrepreneur* and was first suggested by Davies/Ford (1998, p. 13) following Pierre Bourdieu's typological notion of an entrepreneur who embodies various forms of capital (Bourdieu, 1986, p. 241). «*Culturepreneurs*» – so it is assumed – describes an urban protagonist who possesses the ability to mediate and interpret between the areas of culture and of service provision. S/he may then be characterized, first and foremost, as a creative entrepreneur, someone who runs clubs, record shops, fashion shops and other outlets, who closes gaps in the urban with new social, entrepreneurial and spatial practices. Such knowledge- and information-based intermediaries increasingly emerged in the gallery, art and multimedia scene in different European metropolises, foremost in London in the 1990s (Grabher 1998, 2001). Davies/Ford (cf. opt.), characterize a type who, in structural terms, is a communicative provider of transfer services between the sub-systems «business-related services» and «creative scene», and in doing so seems to satisfy a necessary demand (Koppetsch/ Burkart 2002, p. 532).

Second, due to this relatively vague analytical definition, the term *Culturepreneurs* represents an open (re-) search concept. With respect to the current debate on blurring boundaries between economy, culture, knowledge and politics in the urban context, I propose to consider the economic, cultural and spatial practices of the new cultural entrepreneurial agents as testing cases in an urban 'laboratory' situation. In the context of lasting economic crises they might play a decisive role as incubators and attractors for the formation of new creative knowledge milieux (Matthiesen 2004, p. 111). Their creative and innovative business and art practices might combine local skills with creative knowledge and new ideas especially in a de-industrialized and stagnating urban economy such as the one of Berlin.

Debating «Culturepreneurs»: new urban elites or post-modern bohemian underclass?

Following the end of the Kohl era in 1998, a new political beginning of the type wished for by the generation of '68 seemed not to be directly achievable. Germany's new holders of power took as their model Britain's Prime Minister, Tony Blair, and the ways, means and strategies of his «New Labour».

His politics provided the template for the envisioned new beginning of the German Federal Republic. Blair, however, did not carelessly throw overboard the political ideals of the British Labour Party, the party traditionally representing the working class. Rather, he concurrently developed the vision of a new Great Britain with the slogan and buzzword «Cool Britannia», for which he semantically prepared the ground. With the reformulation and reinterpretation, not only of existing social realities but also of those yet to be developed, a forward-looking vision was created, initially on a linguistic level. Carefully timetabled, and often a surprise, deregulation then appeared on the political agenda: neo-liberal realities showed up in the linguistic guise of views of society, which were fit for the future.

Politicians and economic policy-makers thereby assigned a forward-looking role to creative professions in the new cultural economy based on information, creativity, knowledge and innovation. (e.g. Landry, 2000; Leadbeater, 1999) With the slogan 'creative industries' (a marketing concept originating in the culture industry and the broad field of creative work), they promised themselves, at the end of the 90s, the generation of new forms of work, new work places and innovative markets. (Banks et al., 2000)

After a first wave of very optimistic attitudes towards the new leading role of cultural producers at the end of the 1990s, the work and life relations of cultural producers, and their relations to their social and urban situations, have in recent years been the objects of increased critical scrutiny, on both the local and the global levels. This occurred against the background of the stylisation of cultural and generally unpaid or underpaid activities and creative professions. Formerly assumed to be exceptions to wage labour, they served as models of self-determined work in post-fordist society, in order on the one hand to press ahead with the dismantling of state responsibility, and on the other to promote the entrepreneurial self-optimisation of the individual (Verwoert 2003, p. 45). In this respect, the catchword «new entrepreneurship» alludes to individualised marketing strategies and social hardships, but also to skillful alternation between employment office, employment and self-employment structures.

In 2004, there occurred in the European cultural and scientific community a shift of perspective with respect to the relation between self-organised creative work and the politically and economically defined cultural economy. It was the daily experience in different European cities such as Paris, Barcelona, London, Berlin, Munich, Zurich, and Madrid that forced many to rethink their societal as well as individual roles (McRobbie, 2004). Especially cultural producers as well as social geographers, sociologists and cultural scientists began dealing and debating with new economically and socially conditioned mechanisms of exclusion and inclusion.

Today, it is true, the attempt to capitalize on creative work and to bring it under the direct control of the capitalization process that is summarized in the phrase creative industries, has lost much of its public appeal with the flop of the New Economy and the «Ich-AG» (i.e. German for «I Inc») government. But the conversion to a society of self-reliant creative entrepreneurs who successfully market their own obsessions is still underway, only less glamorously than in the mid to end 1990s.

The concept of the *Culturepreneur* has even become a new export good: In the case of Berlin, e.g. the city's public relation agency «Partner für Berlin» makes an effort each year to send a number of entrepreneurial web-, fashion-, and multimedia-designers abroad to represent and market the «New Berlin». The design-oriented branches are important inspirations for the export hit of a young and trendy creative «Berlin», which has helped, as an urban and national label, to create a specific marketing identity for the most diverse creative industries.

Analysing «Culturepreneurs»

In this chapter, I will discuss towards what this social reconstruction, personified by the figure of the *Culturepreneur*, is tending, whom it serves, what it embodies spatially, as well as what effects it has on the constitution of urban scenes. The thesis thus emphasized is: It could be rather important to investigate locations and spatial materials used by the *Culturepreneur* in order to grasp his/her role in the reconstitution, reformation and performance of new social formations, such as scenes in the age of an increasingly individualized and fragmented urban society.

Following Koppetsch/Burkart (2000) and Casey (1995), I claim that, up to now, social diagnosis has ignored systemic changes (in the economy, culture, politics etc.) concerning cultural professions (and their norms, rules, values, practices). Within the framework of a spatially oriented sociology of work, an analysis of new job profiles in the field of knowledge- and culture requires the systematic integration of cultural aspects, new communication and learning strategies, and modes of sociality. The crucial role of the spatial and locational aspects for the formation and establishment of new, and in the beginning insecure start-up business practices, especially under the conditions of the so-called New Cultural Economy, has so far widely been ignored.

Therefore I will analyze first the type *Culturepreneur* as primarily diagnosed by Davies/Ford (op. cit.). I am aiming at extending their notion, and will query if those new professional intermediaries can possibly also be regarded as *space pioneers*.

The extent to which their appearance in urban areas can specifically be explained by involving geographic as well as social space is examined. I will clarify which abilities are attributed to the *Culturepreneurs*, what kinds of agencies they require – or create for themselves – in order to build up networks, to arrange meetings, and to establish laboratory places where new products can be tested, and where experience and knowledge may be shared. Which urban locations do they need? Will they create their own locations and landscapes in the absence of suitable existing ones? How do they communicate, perform, and present themselves beyond the traditional settings offered by employment agencies, trade or art fairs and corporate associations?

Second, the aspect of performativity, as well as the performative role of the *Culturepreneurs* in an urban context and in the development of cultural clusters, called *local cultural industries*[2], has not yet been a subject of discussion. Accumulation of cultural facilities and «cultural capital» in one place makes – as often assumed – a positive impact on the site policy of «placeless» service economies. Particularly, the *new creative workers* active in the «business-related services» sector expect and require such social and creative milieus for their professional activities (Helbrecht, 1998).

This chapter investigates interpretations contributed by this cultural and entrepreneurial type to urban conditions, while addressing the cultural-economic modernization status of the new German capital Berlin. Contrarily to Blum (2001, p. 7), who concludes that «the body of literature and research on cities seems to be silent on the questions of scenes», theoretical discourses will be integrated into empirical results, in a first step to overcome this silence.

To put it in more abstract terms: If the integrative machine of «the city» no longer functions comprehensively, which visible and invisible social micro-formations will appear in an urban society? Which of these will take the place of the traditional, formal and, concomitantly, democratically checkable forms of work and mechanisms of integration?

The following section (2.) will start with key theoretical prerequisites to reconsider the formulation of a «cultural turn» in the spatial sciences. First quantitative structural data demonstrate the ups and downs of new design-intensive services. The ensuing discussion about the convergence of urban agendas and its roots in popular culture and new products is closely linked to the emergence of new professional biographical careers (3.). It is followed by the presentation of approaches to the social and spatial establishment of new forms of urban communalization (4.). An exemplary implementation of this diagnosis with empiric field material will be used to explore the initial question further, so as to present first findings relating to possible tactics and strategies (de Certeau, 1988), developed with content-analytical evaluation methods using the capital Berlin (5./6.). A first interpretative conclusion will sum up the empirically based scene- and place-related practices of the entrepreneurial actors examined in an ongoing research process (7.). Hence, what follows is an analysis of the general conditions that may grant this cultural and entrepreneurial type a decisive role in a potential formation of new approaches to and methods of dealing with urban, economic and cultural modernization processes as well as landscape transformations, which are presently taking place in Berlin in an exemplarily manner.

CULTURAL TURN AND SPACING

Within the framework of a theoretically formulated socio-geographical «cultural turn», cultural codes, meanings of space and processes of evaluation, as well as devaluation, have been shifted to center stage in recent socio-spatial analysis (Cook *et al.*, 2001 for an overview; Philo, 2000, p. 27). Taken to the extreme, this approach is criticized for practicing an «anything goes» eclecticism with no sense of political project, neglecting the political economy, abrogating all responsibility for asking questions concerning value, quality and truth (Thrift, 2000, pp. 2-3). The German debate seems to focus on a «dosed cultural turn» (Matthiesen, 2002a, p. 26). Within this perspective, the focus of spatial analysis has shifted towards a relational understanding of cultural codes and persistent physical spaces, materialized spatial constructions and culturally coded identities.

Spatially relevant action patterns, networks of specific actions and the formulation of institutions are always coordinated and mediated by symbols, processes of communications and modes of governance. Matthiesen (2000a, p. 28) concludes epistemologically, with reference to Appadurai-Breckenridge *et al* (2002), that a uni-directional space- and world-relation is principally not possible. This is often suggested by key terms used in urban planning processes[2]. Moreover, the renaissance of «soft» structures, the growing knowledge-based economies, the hybridisation of cultures through migration, the interdependence of technically-mediated global economies as well as the competition of cities through images on a trans-national level, demonstrate that these new relations between economy and culture follow the model of structural homologies and not the uni-directional Marxist approach of basis and superstructure. The boundaries between social reality and representations of this reality seem to collapse. They require new theoretically and methodologically based approaches.

The debate on the «cultural turn» has stimulated new conceptual linkages between the formation and readability of cultural landscapes, the culturalization of existing urban fabrics as well as the socially-divided norms about the organization and evaluation of the physical space through the view of different individuals, scenes, groups, milieus, and institutions. Haber concludes that «Landschaft» (Germ. for «landscape»; see chapter - Olwig) takes a metaphoric role as a holistic mediator between subject and object, as well as the individual and the individualized society (2001, p. 21). Following this division, socio-cultural landscapes in an urban system consist of numerous recurring patterns embedded in a background-matrix. According to Forman/Godron (1986), we can epistemologically approach socio-spatial patches, places, corridors, nets, mosaics and social micro-formations. These terms can be understood as an expression of a social *gestalt*, which is visible or detectable on a micro-meso level, thus representing and referring to a larger system with abolished boundaries (Giddens, 1990).

Due to the theoretical efforts of German urban sociologist Martina Löw (2001) new systematically- and theoretically-developed tools can be applied to the empirical analysis of the diverse social formations of new agents and their place-making strategies in an urban system such as Berlin. A central prerequisite for such an approach is that categories of space, which have in the past been regarded as being only marginally relevant, now be brought to centre stage. Here, space is to be understood as the result of an act of synthesis based on the specific strategies and tactics of individual protagonists. The term «spacing» describes the active process by which an individual relationally orders social goods and bodies (Löw, 2001, p. 158). Hence, space constitutes itself as a process through synthesis of these social goods and bodies, by means of perception, memory and feeling. In the post-industrial city, individual differentiation strategies are symbolically and culturally formulated. The socio-spatial structure expresses itself increasingly strongly in local politics through which individuals, not only create symbolic difference, but also attempt to arouse attention through positioning tactics anchored in the location.

BLURRING THE BOUNDARIES BETWEEN THE SUBSYSTEMS 'ECONOMY' AND 'CULTURE'

Context and first findings

Berlin today lacks a substantial industrial or services-based economy. Since 1990, the city's potential has been in its knowledge and education basis with various universities, polytechnics and research institutes, and in its cultural attractiveness. Official authorities have heavily promoted new business and knowledge fields, such as media and film production, as well as biotechnology (Matthiesen *et al.,* 2004).

In the last year, one of the most significant professional groups, the design sector, has grown substantially without direct support or financial aid by the city administration. Recent Berlin-based surveys indicate the following situation: In July and August 2003, the International Centre for Design in Berlin (IDZ)[3] (Internationales Design Zentrum, 2003) evaluated the quantitative structure of Berlin's design scene. Fashions, web, architectural, print as well as media design are included in this group. Of a total of 1153 listed enterprises, 18.2% had been founded in the last three years, 31.5% in the last ten. 84.4% of all enterprises in 2003 employed 1-5 persons. Almost 50% of the total are working in «different professional» fields. So, in order to remain on the market, they had to leave their trained profession and move over to relatively new service fields. Around 39% of all enterprises have regional, 34% national clients, only 7% has international clients, and 12% offer no regional specification (8% did not respond).

The crash of the so-called New Economy in 2001/2 had a tremendous impact on the development of this professional group. In 2000 and 2001, 186 new companies were founded in Berlin's design sector; in 2003, their number decreased to only 48. Many employees lost their jobs and had to move in new directions (Krätke, 2002, p. 139). The number of media and communication designers grew by 10% between 1994-1998 in Berlin, compared to 6% in all of Germany (Krätke, op. cit.). These key figures demonstrate structural changes about a relatively new and in the last years growing professional group growing fast in the past few years.

Grésillion (2004, p. 197) stresses that these new professionals are agents of a «global society of consumption» and practicing a so far unexpected model of a politically formulated New Centre coming from the art scenes. Wiedemeyer/ Friedrich (2002, p. 57 ff) emphasize the obvious fact that in Germany the labour market for artists and their related professional fields is considered to be one of the most dynamic ones. They estimate that between 2000 and 2010, the number of employed artists will double and reach a volume of 460.000 (op. cit.). Alongside the growing importance of evaluating, communicating and transferring knowledge and information, lifetime investments in education, knowledge and cultural capital seem to be, first of all, an individual strategy in engaging with these mega trends.

To summarize, these indicators first demonstrate the key role of new labour profiles for the extremely flexible and unstable job and service sector. Second, in spite of persistent national economic stagnation, various researchers and politicians expect a growing number of professionals and companies to invent new products for new markets.

Convergence of urban agendas in European cities

Davies/Ford (1998) were among the first critics along with Urry/Lash (1990), and Scott (2000) to investigate and discuss the convergence of economic, political and cultural agendas in London and other European cities in the late 1990s. Much of this activity has been a direct result of organisational and structural shifts occurring in the corporate sector, principally the shift from centralised hierarchical structures to flat, networked forms of organisation. In order to understand this process, it is becoming increasingly necessary to look at how networks and 'new' economies are being formed, accessed and utilised, and where they converge, interact and disperse. In the journal 'Art Capital', Davies/Ford (1998) identified the surge to merge culture with the economy as a key factor in London's bid to consolidate its position as the European centre of the global financial services industry.

'Culture' was part of the marketing mix that, within the context of the European Union (EU), kept London ahead of its competitors, particularly Frankfurt on Main in Germany. This can be traced back to the UK's exit from the Exchange Rate Mechanism (ERM) in 1992 and a range of economic initiatives aimed at attracting inward investment, or Foreign Direct Investment (FDI). During this period, the UK accounted for 40 per cent of Japanese, US and Asian investment in the EU. Contrary to the media spectacle of 'Cool Britannia', it was the need to attract FDI, combined with the co-ordinates of a New Economy based on knowledge and services, that underpinned London's spectacular emergence as the 'coolest city on the planet' in the late 1990s.

These developments opened up the field in which artists were encouraged to operate and many took advantage of emerging cross-promotional opportunities to redefine and diversify their practice. As a result, identities based on terms such as artist, curator, critic and gallerist came under increasing pressure as the range of activities and professional services in these areas expanded.

In retrospect the cultural requirements of the New Economy resulted in the emergence of 'culture brokers' - intermediaries who sold services and traded knowledge and culture to a variety of clients outside the gallery and art system, from advertising companies and property developers to restaurateurs and upmarket retail outlets. Their bridging role to communities on the 'cutting edge' of culture provided the local knowledge, information and association vital to businesses pursuing 'multi-local' global strategies.

Angela McRobbie first critically analysed the situation of selected creative industries protagonists in London boroughs and their socially disintegrated and fragile professional and living situations, against the background of popular culture (McRobbie, 1999, 2002). Artists organise their own exhibitions or open their own shops. Critics assess such activities as commercial strategies – which they doubtlessly are. But they are also opportunities of getting over unemployment and starting an activity that is perceived by the public. Thus, cultural protagonists are positioning themselves both in the do-it-yourself tradition of punk and in the neo-liberal corporate culture. This marks also a significant shift in the notion of «culture».

«Culture»: From sub- and high to pop culture

Since Tom Holert's and Mark Terkessidis's pop standard work «Mainstream der Minderheiten», (German for «Mainstream of the minorities») in 1996, the clear distinction between sub- and pop culture has been massively questioned. The authors describe the 1990s as having been marked by fast shifting and recoding processes between mainstream and underground, resulting in the formation of a hybrid mainstream consisting of minorities. The class- or stratum-specific model is replaced by the scene-model where social networks with their pertaining symbol worlds can be almost freely chosen. New everyday-sociological categories, such as styles, symbols, tastes and aesthetics, are gaining relevance as means of distinction. Difference is the mainspring of post-modern consumption. The tactics and practices of demarcation (from official culture) used to be genuinely linked to the underground. However, culture industry becomes interested, detects even the most subtle refusal and re-evaluation, and takes advantage of such. This happened to the *punk*, *grunge* and partly also to the *rave* and *techno* scenes: Pop is described as an ever-regenerating potential of social and aesthetic attitudes.

The relation between pop and high culture is similar. The elitist marginalisation strategies of bourgeois high culture (to which the subculture elitisms had structurally adjusted) have been undermined: in order to appear top educated today, one does not only go to the theatre or opera but proves his/her cultural competence also in the field of popular culture.

Particularly inconsistent with the traditional high culture concept are the new creative scenes of cultural agents romping around in electronic media, design, and advertising, but also in the music-, fashion- and clubscapes. Most of the cultural products of these actors meet the highest aesthetic demands, which suggests that highly developed techniques are adopted and made into a new independent system of cultural life. The combination of regularly needed new trends, new products with fresh financial investments and intensive humanpower represents an urban-based speeded-up cultural and media economy.

When producing intertwining goods between the economic-service sector and the cultural sphere, speed and time have tremendous but yet unclear impacts on the organisational processes of such production flows. In the context of this first diagnosis, it is not detectable how new informal linkages between various agents and institutions are possibly woven between corporate companies, the city's administration and new, often temporally organized, networks of cultural producers. Furthermore, the ways in which this speeded-up economy affects the organization of the producers' social living worlds are by no means clear.

From the perspective of the professional situation of the producers, the term *patchwork biographies* only insufficiently covers the wide range of their often unstable professional careers (e.g. *freelancer*) with sharp breaks, insecure life paths and an orientation towards individual entrepreneurialism: a «hot» exhaustive production phase is often followed by a «cold» phase. This jobless phase increases the pressure to personally remain in the market, by holding several jobs at the same time. Networking on professional alliances with people who are at the same time friends and rivals seems to be not only a self-evident and welcome social practice, but also an existential necessity for security. It is not yet known whether new flexible and network-like alliances can be considered an informal institutional answer to the ongoing process of the rise of new subjects of cultural individualisation in the context of the dismantling of the welfare state (Wittel, 2001; McRobbie, 2004).

PROCESSES OF INDIVIDUALISATION AND *'SCENIFICATIO'*: ON THE CONSTITUTION OF URBAN SPACE

Experiencing the urban

An aesthetic of the urban and an atmosphere of the possible create urgently needed fertile soil for the new, creative entrepreneurs, many of whom have planted in it the seed of their own business or urban dreams. This brings a greater emphasis on subjective experience. Although image strategies provide the visual codes and spatial images in relation to which one positions oneself, such images only have a justification for existence when there is a correspondence with workable spatial-sensual contexts of experience.

The politics of the British, as well as German, «New Centre», starting first in 1996 with Tony Blair and in 1998 with Gerhard Schroeder in Germany, has a visual effect on the development of this new type of cultural entrepreneur. The addressees of the politics (and image politics) mentioned above, are representative of a de-structured urban society whose reality is not only extremely individualised, but also equally greatly ambivalent. As a result of numerous unpredictabilities, lost communal reliability, and an alleged multi-optionality with regard to life choices, individuals are required to make a series of new decisions about their behaviour in order to situate themselves in urban society. The spatial location of playful experimentation with this demand for individualisation is the city: the city is seen as the laboratory for one's own ideas, irrespective of the fact that the individual protagonists are subject to new patterns of flexibilisation and processes of social disintegration, which can best be absorbed in the urban.

Individualised entrepreneurial existence strategies, however, are socio-politically positively coded, and have been at least since 1998. Here, independence demanded by politics, and the gradual exclusion from the social security system, were gallantly whitewashed by the choice of language. The result was a politics of linguistic images and the redefinition of symbols. Images and symbols had to be found which proclaimed realities, rather than possibilities, ideas rather than delivery, attitudes rather than events. Following the German parliamentary election of 1998, Gabriele Fischer, editor-in-chief of the magazine *Econy* (sic!), suggested that the presentation of business as an adventure might be the formula to re-ignite desire for the project «work», and proclaim the realisation of daily existence, not as a burden and daily chore, but rather as a source of fun and a personal reality option. This type of politics addresses people «who want to break free, who want to do something, who still see business as an adventure and are not always complaining about bureaucracy and the burden of taxes» (Fischer, 1998, p. 1 ff).

The city as an adventure playground

Retrospectively, Gerhard Schroeder's politics of image making – in contrast to its declared future orientation - greatly romanticises the image of a self-reliant, pioneer-like entrepreneurship in the adventure playground of the city. The city organism, on the one hand, thus appears as a potentially chaotic, open, but at the same time«cool» territory that provides the ideal development opportunities required by the *Culturepreneurs*. On the other hand, the logic of the adventure playground means, subliminally, that as part of this individual entrepreneurial campaign, real possible risks should not be judged to be existential threats, but should rather be understood as new opportunities for orientation, that is to say, as opportunities for acquiring personal knowledge for future operations. As expressed by Schroeder in his 1998 statement of government policy, the «social net» of the welfare state must «become a trampoline».

The counterpart of these ideas in images can be rediscovered, in modified form, in the marketing strategies of cities. The city is increasingly establishing itself on media image levels, constructed according to imaginary models (for example, historical Middle Ag, Tuscan or Mediterranean, etc.) and fixed typologies whose socially integrative relationship is difficult to discern. In the municipal context, image strategies are increasingly geared towards a group which is young, dynamic, happy to make decisions and willing to consume, and which in turn tries to correspond to the ideal type represented in these marketing strategies for the urban. In this way, specific urban articulations, for example in Vienna a mid-90s kicking Drum 'n' Bass scene, in London hip and trendy fashion makers, and in Berlin young and cool multimedia and designers, became known to a wider audience. The cities, in turn, see in this clientele, whose self-image comes very close to that of a company, the «I Inc.», the opportunity for an expansion of the business-related service sector, for economic prosperity and for purpose of profit making.

One positive, yet ironic interpretation of this emerging milieus, could assign to these new type of entrepreneur a much needed function in the economic sector of the city as the creator of bridges and systems of communication between the two subsystems of economy and culture (Bude, 2001, p. 9). At these hubs of communication, whose physical equivalents are club events, gallery openings and start-up opening parties, questions concerning the modernisation of the city are addressed anew. It is not the self-presentation and self-celebration of the individual, which should be awarded as principal significance of these patently performative events. The«places of the *Culturepreneurs»* are, rather, a platform for social interaction and transfer, than a permanent and purely economic-structured place; platforms on which, using urban materials, new relationships can be tried out. The question for the present urban-planning and urban-development making of places in the city in times of significant and generally accepted transformation processes is inevitably linked with the question concerning whom such places shall be made or changed for.

Modes of social and spatial re-embedding

Socially disintegrated, the urban individual finds him/herself exposed to a large variety of action alternatives and decisions that can facilitate societal place-making for him/her (Beck, 2001, p. 17 ff). *Dis-embedded* from known socio-structural safety-nets, such as church, trade unions and codified associations and increasingly also family ties, the individual has to decide on some social contexts – such as milieux or scenes - that ensure timely flexible social integration opportunities.

However, structural modes of «post-traditional communities», noted in this process (Baumann, 1995, p. 19), are very much unclear. Post-traditional communities differ from «settled and established communities» (Hitzler/Pfadenhauer, 1998, p. 78) in that memberships can be revoked at any time since they are largely free. In addition, another difference according to Baumann (op. cit.) is that they create an «imagined or aesthetic community» which provides the short-term illusion of being coherent, in terms of forming opinions as to what is right and relevant. It has authority, as long as it is assigned authority, since it has only little institutional sanction potential. The power aspect, posited by Baumann and Hitzler/Pfadenhauer, is based on the potential of persuasion, on the per definition «voluntary emotional bonds of the actors conceiving themselves as members». However, from an analytical view, there are also obscurities with regard to participants of the «persuasion debate», such as stated by Baumann (op. cit.), particularly on the question for the *How* and the critical examination of individual contexts. *How* and *where* does the establishing core of the communalisation processes form? Or, more directly put: How is urban space constituted nowadays? Who literally works towards its constitution?

Initially, it appears to be less plausible to identify purely aesthetic and visible surface phenomena (clothing, outfit, etc.) as modes of integration and motives of the desire to participate in scene-formation processes. While one may surely consider this sub-aspect, questions arise with regard to the deep structures and subjective motivations of actors integrated into specific social, political and mental formations. My findings suggest that mere participation and assimilation in a collective during one of the much-quoted techno or rave events can*not* – in analytical terms – already manifest the core of processes of scenification[4].

Performativity and spacing. On the constitution of scenes

More recent approaches, however, explicitly explore the rank of space and show that space-establishing processes are progressively more complex. Theoretically Löw (2001) pointed out that such processes are «brought about in acting by a structured arrangement of social goods and people in places.» According to Löw (op. cit., p. 204), «objects and people are arranged synthetically and relationally». Löw posit that spaces are not always visible formations but can also be materially perceived. Accordingly, spaces are ascribed a potentiality that is characterised by the term «atmosphere» (Böhme, 1995, p. 24 ff). Following Blum,

> «…the element of theatricality integral to the scene marks the importance of its site as an occasion for seeing; the scene is an occasion for seeing and being seen and so, for doing seeing and being seen. But to mention its occasioned character is to bring time as well as space to the grammar of scene. For if the scene is a site, a space for seeing and being seen, its occasioned character marks it as the site whose engagement is punctuated temporally as if it were a ceremony (Blum, 2001, p. 14)»

The constructivist approach of Blum (op. cf.) and Löw (op. cf.) argues for a changed relationship between body and space, the individual and the temporal collective at the centre of attention. The task at hand is to address the experiences and emotions, which are intrinsic in physical bodies and interact with built-up space. After Wigley (1998, p. 18), entering a building (and a city) means to experience and encounter its atmosphere. Atmospheres are created purposefully and, hence, in a societal manner. According to Böhme (op. cit.) they are a reality that is shared by the perceiver and the perceived. Blum (2001, p. 17) suggested that «performance is transgressive in its very potential to create exposure or humiliation even in its most mundane shape such as calling a spectator to perform (…) in a way that dissolves the border between audience and performer.»

If a potentiality is ascribed to spaces, the spaces are bodily and physical actors, Due to their radiation, which was insufficiently considered and analysed in the past, new processes for the establishment of new forms of communalisation can be detected. The truth is that the distinctive place is less – albeit also – (purely) physical than a variable that is deliberately chosen by individuals because it is suitable for staging compatible opportunities for group formation processes.

Bette (1999) sees the increased presence e.g. of skateboarders, but also of other phenomena, such as general sports activities, artistic actions, and performances in shopping and consumption areas in inner cities, as existential attempts of individuals. They are especially aimed at attaining a new affective-bodily experienceability and an enhancement of the intensity of their sensitivity not in just any but in deliberately chosen places[5]. Bette takes up current phenomena and explains that such phenomena only come to bear with a constructed atmospheric net, a space image, and, following Blum (2001, p. 18), a «collective representation». This net generates and simultaneously radiates meaning, but experiences its purposes only when the actors integrate their bodies and their preposition to experience this situation into the temporary power field.

This brief and cursory overview of current work in the field of diagnostic modes of communalisation and scenefication in an urban environment shows that, after the emergence of new professional and activity fields in cultural production there are only very limited and inadmissibly reduced queries on the changes of meaning, also in relation to urbanity. Hence, what follows is a further analysis of the negotiating places where this type is responsible for the re-establishment and re-formation of scene-like communities, in the age of an increasingly individualised and fragmented society (Hitzler, Bucher, Niederbacher, 2001).

GENERATION BERLIN?

Widely noticed but rarely integrated by municipal powers in communal politics, the number of private exhibition rooms and clubs in Berlin has multiplied in the 90s, particularly in the latter half of the 90s. Never the less, the do-it-yourself attitude of the organizers had its impact on the municipality: till the mid 90s, the illegal character of many creative initiatives was ignored or silently accepted. This «laissez faire» attitude was swept away by investment capital and reconstruction activities. The fact that the efforts of these creative initiatives have their roots in youth culture suggests a new authenticity, which is also an indication of a self-regulated variety. So far, the protagonists of this youth culture have found themselves constantly at loggerheads with the municipal authorities.

Berlin involuntarily supplied these cultural initiatives with open space, empty buildings and also unclear planning situations. Due to the ongoing process of urban unification, existing open spaces provided the existential base for those activities. These cultural projects of the 90s were confronted with the practices of many initiatives with ideological roots in the 70s and 80s. 30 to 40 years ago, Berlin's International Building Exhibition (IBA) and their efforts worked towards establishing and strengthening the local level, the Kiez[6], as the distinct form of a geographically based and essentialist locality. As they sometimes cooperate, sometimes struggle against each other a conflict is diagnosed between two generations with distinct and significant differences in their spatial practices.

By using the case of Berlin I observe new tactics in spatial practices by new urban protagonists, the *Culturepreneurs*. While Mc Robbies speaks of processes of spatial disintegration, for example, of centres of creativity (Mc Robbie, 2002, p. 479) the case study that follows demonstrates new societal formations and emplacement strategies invented by *Culturepreneurs*. This analysis provides inside knowledge on current social formation structures. The following urban platforms deal with action in the context of the tense relationship between basic neo-liberal existential conditions and the desire for an economic and artistic realisation of one's own ideas. On a methodological level, this simultaneity and complexity of observed processes refer all the more to context-sensitive approaches and a view of transformation scenes which suggests that it is appropriate to examine the emergence of the new also as something novel (Matthiesen, 2002b, p. 130)[7].

A preliminary methodological remark: individual cases mentioned below aim at generating themes, categories and narratives from sequences of guideline-supported and semi-standardised interviews. What will come to the fore are, above all, living world-related aspects as well as situational and socio-spatial ascriptions that provide a superior-level explanatory basis for certain actors, their associated professional groups and networks. Those levels of meanings are reconstructed, i.e. developed from the statements in the interview material, subjected to comparative reviews and contrasted. Some first generalisations of professional and biographic transformation situations and their spatialisation thus become possible. The latter do not provide information about an individual case, but about the specific milieus, scenes, and social arenas, institutional local, regional and supra-regional intertwining and structural situations, which are articulated in the sequences of the case. Ethnographic field observations generate further sources of information.

PROJECTS, PLACES, AND PRACTICES OF *CULTUREPRENEURS*

Case Study 1: Launching an entrepreneurial project

Preparing the place

The first case study deals with three men, aged 27, 32 and 35, two of them studied graphic design in Cologne until 2000. They have worked there during their studies in different offices and agencies, and acquired additional experience after graduating, as employees with far-reaching competences and tasks in other – also international – agencies. In late 2001, they moved from Cologne to Berlin, searched for office space in Prenzlauer Berg borough, where they found a suitable office. It was a floor space of approx. 145 sqm and was a disused shop near Helmholtzplatz. The rent was quite cheap, they redecorated the rooms themselves, brought their equipment, and 'organised' table boards. An Internet connection was established via a free hanging cable from their window across the courtyard to another office next door. Work commenced with an enormous party.

> Field record: «We are sitting in the anteroom and they explain to me some of their projects, also some of which are only in the stage of plannung. It is not really peaceful, the anteroom has to be cleared since a friendly gallery owner and her artist plus some

works from Bremen are due to arrive tonight. The anteroom facing the street – where work is done at four large desks during daytime – is emptied. The next-door DJ is setting up his sound and light equipment in the doorframe (the light gear includes a multi-mirror disco sphere and a video projector), some beer crates are lugged in.

Half an hour later, the room is empty, the curator appears and starts placing the objects, frames are put up, carpets unrolled on the floor, flyers and information leaflets put on display. There is no formal opening. At some point the whole room is filled with people. At about 11:00 pm the room is brimming over, smoke and electronic music waves across the heads of the visitors, some of whom are dancing. More people pass by and crane interested heads into the place or push off. Colleagues, competitors and critics arrive, while most discussions focus on potential orders, past jobs and safe contacts.

At about 1:00 am, two of the three office owners stand outside, in front of the door to exchange views with colleagues from another agency. I learn that two of the three owners have registered as unemployment with the labour office, and use another person, namely the third owner, as a stooge to bill their services, and that the two former owners are currently trying to organise start-up capital out of their phase of unemployment. The field discussion was started by asking 'are you also going about it like this...?', which was casually answered in the affirmative (End of record, documented on 23-5-2003 by the author).

Forming identities: «Universal dilettanti»

Their identity-creating work is rooted in their training as graphic designers at the University of Applied Sciences so that they may be called – in the widest sense – design-intensive symbol producers. They define the specifications of their production with the term «holistic designs», which for them implies necessarily high design standards as well as an artistic self-image in the performance of their activities. This specification shows, on the one hand, artistic motives, on the other it is unspecific, adjustable and extremely variable in terms of content.

Their project-based studies at the University of Applied Sciences were characterised by open structures or 'quasi pre-programmed independence'. The project studies trained them to become «universal dilettanti». These self-ascriptions show learned but also unquestioned flexibility aimed at attaining the necessary and professionally desired expressivity, as well as absolute professional, but also personal control of the content of their products.

Conditioned by the socialisation patterns of their educations, their transition to their pure work life appears to be basically successively prepared from their living world. Actually, there is no well-defined entrance into work life; education and work life were rather entwined over several years. The continuation of those entwined phases systematically manifests itself in the conceptualisation of their business: the latter will be established, besides its thematic openness, also in socio-organisational terms, as an interaction platform and docking-stepping grounds station for other actors (friends, acquainted and professionally associated actors etc.), that are deemed of project cooperations. Consequentially «working in a team» entails a professional integration of like-minded friends, partners and even lifetime companions. Consistently, the name of the «Greige» office does not cite the names of the owners, as is customary in Germany, e.g. Springer & Jacoby. The name «Greige», a reference to a colour between grey and beige as coined by Le Corbusier, refers not to persons but to the socio-organisational idea of interdisciplinary work in network structures.

Work in networks is structured systemically, where every actor from different European cities who temporarily collaborates on a project, contributes his or her skills to the current work. This organisational structure can highly, swiftly and flexibly adjusted to any external requirements: Thus, new enquiries and orders may be addressed to within a few hours or days, by putting together appropriate teams. A suitable team can be presented to third-party clients not only as a quantitatively large but also perfectly suited design office.

This organisational model consistently combines and links work and private spheres. Strictly speaking, there are no classical work time models and time structure models that find application in all situations. The previously separated living world spheres of «work» and «leisure» are defined according to specific order and employment situations. The organisational structures of actors in the field of symbol-intensive service provision swiftly point to – as it is documented in the above record – hazardous subsistence conditions of urban cultural-economic transformation structures, but also to strategic responses of individual precarious situations. Hence, future course will focus on questions – in the sense of an intentional position – about tactics and strategies that can be derived from the self-ascriptions of actors and that are developed in times of extremely low competition (due to hardly any order intake).

Special emphasis is placed, on the one hand, on free design competition, networking and integration approaches, as well as on cooperation with associated offices. On the other hand, we can hypothesise from our observations that micro-spatial strategies are used to subject immobile and ostensibly clearly programmed office space to various sorts of temporary change and re-programming. However, the following is less an artistic or effectiveness-related evaluation of such micro-spatial policies, rather an investigation of the variable range of strategic approaches adopted by expressiveness-geared professional groups as they offensively reacting to difficult subsistence conditions in times of economic crisis and structural upheaval. Place matters!

Such internal orientation and (re-)structuring of places shows a sample of elements of responses to the extreme structural crisis and scarcity conditions of Berlin. Yet, besides their organisation as a flexible supply-oriented platform with extensions in many other European cities, I have also identified approaches to a development of creative demand options. The necessary organisationally competitive character of the 'Greige' enterprise highlights communicative strategies in order to make proactive use of the micro-location 'office', in the sense of *place-making*, as a hub for fluid social communalisation and cultural scenes.

Playing (with) the places

Culturepreneurs' locations are part of a highly individual and, at the same time, playful practice of (attracting) attention. In order to register locations in the minds of other people, a specific policy of location and scene is necessary, which renegotiates (a sense of) cultural belonging. *Greige*, for example, may be the meeting place for an open, but clearly defined, group of friends, colleagues, rivals, for the interested and for the curious. Its access and perception are guided by policies which displays similarities to those of a club. However, the well- known selection mechanisms of a club – i.e. bouncers turning people away at the door – does, though, take a rather subtle form in the case of *Greige*. A variety of media, such as word of mouth recommendation, or mailing lists and flyers ensure that information on forthcoming events, exhibition openings, or even new products, reaches a specific target group. Apart from this information policy, however, efforts are also made to ensure that the location *Greige* occasionally recedes into oblivion. For months on end, nothing happens; no events are organised, partly, because there are other matters to be attended to.

In the case of *Greige,* we can see that a game is being played with visitors, camouflaging the location, and then returning it to public consciousness, at a later date. *Greige* works without an annual plan and announces their/its art exhibitions, at short notice, by sending invitations via e-mail lists, above all to selected friends and interested members of the wider, also other European-oriented Berlin art scene. The header on the e-mail indicates (or fails to indicate) membership to what has thus ostensibly become a scene and is the criterion for inclusion or exclusion in a social formation about which no one bothers to talk openly.

This at first surprising and seemingly contradictory strategy of hiding is a sort of behaviour, which evokes memories of the old socialist mentality with regard to the service industry: the customer is not king, and business apparently doesn't matter. This strategy is also employed outwardly: To the outside world, the appearance of the location *Greige* offers no indication of what events take place inside. Only insiders and those with local knowledge perceive it as a place where events and performances take place; only they can read the local urban landscape. In positioning itself in urban space, by means of this policy of hide – and-seek, *Greige* creates not only social difference, but also keeps the broad masses at a distance.

When eager searchers do nonetheless find the location, another subtle differentiation criterion is brought into play. At the parties which take place, for example, after openings, present guests are offered a variety of identification patterns by means of art exhibited and electronic music played. It is the assignment to these cultural-symbolic products – based on the extent to which the performance can be experienced and interpreted – which, in the first place, makes a memorable participation in the event possible. This is where the subtle exclusion strategy lies: No one is refused entry to the location; indeed it is rather the case that anyone may be admitted, but only a few are integrated. This integration is also a challenge to secure membership on a permanent basis, the changeable character of the location guarantees, in the first instance, that no trend is created, that no financial dependencies arise, and that commerce does not hinder the creative enterprise. It is this act of maintaining a balance of permanent change in the differentiation criteria, an avoidance of pure commercialisation, and the employment of hiding strategies, which ensures the survival of this location and its protagonists, for some time. If they were to position themselves as an open counterpart to existing cultural and social currents (the «underground» model), they would immediately be culturally chewed up by the urban trend machine and financially destroyed, as indeed they would be, too, if they adopted the «mainstream» model (as was the case in Berlin for Hackesche Höfe or other locations).

Their interest in location, and in what location expresses, indicates, for one thing, pleasure in the local coding game. Pleasure, however, comes up against the necessity of dressing the location in a specific narrative of location-symbolism in order to be perceived at all. In this game with the significant, locations are, for this reason, battlefields of symbolic landscapes in the post-industrial city. Subtle tactics of social positioning can be observed at (and in dealing with) these locations. Even places in relatively established housing areas display heterotopic characteristics. They can no longer be categorised as underground or mainstream, as would have been perfectly possible a few years ago: those who operate and play at the locations have achieved a degree of reflection, which makes it possible for them to employ emplacement tactics which work with and play in economic and cultural terms with social Utopias, with alternative blueprints. They make use of traditional standards of Bohemianism, but are, by reason of this very act of adaptation and by reason of their understanding of the *Zeitgeist*, pop-revolutionaries and, so, responsible for post-urban transformation processes.

Case Study 2: Exploring and designing Berlin

Ethnographers and storytellers of the urban

Our interview partners (one female, one male) are 26 and 27 years old and hail from Luzern (Switzerland). They received training as graphic designers in Switzerland. In 1997/8 they relocated to Berlin where they worked as interns in several agencies, took on the *Art Direction* of a magazine and did mainly freelance graphic work. Since January 2002 they have worked together in a disused shop in Friedrichshain, borough Kreuzberg-Friedrichshain. They have been integrated into a residential quarter management project titled «Boxion».

Their identity-creating work is rooted in a wide range of creative design production, mainly in the print media but increasingly in the Internet sphere. Besides smaller orders, they were given the opportunity of designing the magazine 'Berliner' in 2000. Three issues of this high-quality magazine were published, before it was discontinued. It was a medium in which – *nomen est omen* – Berlin was re-discovered. The following is a closer review of the identificatory content in the work on a Berlin magazine by the two Swiss partners mentioned above.

What is striking is their approach to 'move something with a company formation'. This moment of movement and moving in space presupposes a space that is not pre-moulded, or better, it presupposes 'space' per se. This sense construction of a space, that is not pre-moulded in their view, forces the actors to develop strategies of self-assertion in space, to quasi discover their own territory, and to symbolically occupy and re-code it. The counter-horizon of Berlin, formulated and stylised in the process as a terra incognita or a NO LAND consists of the morbid charm of the former workers' borough Friedrichshain, cultural artefacts in the form of East-German residents, their hidden leftover stocks of cultural knowledge, socialist practices and behaviours. Thus, this self-ascription of the actors shows a moment of ethnographic significance. The two actors conceive themselves as strangers in the city of Berlin, and basically take on the role of ethnographers, via which they re-define and re-evaluate the social relationship between insider and outsider, old and new, in and out.

They use a variety of attributes to describe a romantic situation of Berliners, an almost extraterrestrial situation, a spatial peculiarity that is hard to match. The symbolic space of Berlin appears to the actors, who originally hail from Switzerland, to be a project, a space for movement, that seems to be, for the most part, suited to their current living world, and their professional and social situation.

«Being Berliner»

This sense-making role, however, does not only exhibit exclusively self-referential aspects, but also includes clear signal patterns of a network-like sociality: Both actors develop a product, a magazine, titled «Berliner», which represents, on a graphical level, the disparate cartography of how the two protagonists perceive this city. In other words, this product bundles social and psycho-geographical orientation knowledge that is distributed to a fluid community of temporarily like-minded people.

This process imparts an image of a new location in Friedrichshain, a «scene and pub borough» that has been prospering for years, as a mobile event space. The 'mood level' and new borough initiatives and siting practices by 'culture management agencies' (such as the «Spielfeld» agency[8]) are accompanied by very concrete structural changes at the 'business structure' level. Thus, the semantic, graphic and event-oriented special expertise is produced, imparted, distributed and supplied to a broad clientele by 'scene' experts. It is at this intersection that the ability to recognise socio-spatial potentials and their economic utilisation is combined with entrepreneurial philosophy. The company name chosen by the actors, 'Substrate', is a program, since it commodifies their entrepreneurial cultural practice and hence, represents their living world identity.

This construction and situational self-ascription forms the bed of the bohème-like marginal position (that is also related to a cliché) from where the two Swiss can position themselves as artists and make professional use of their situation. Hence, the obvious positioning and self-ascription fits in with the spatially conditioned location perception in Friedrichshain so that the – in their view – disparate transformation material (old workers vs. new designers, anonymity vs. socialistically idealised practices, etc.) permits the recognition of a stimulation substrate which, in turn, is reflected in the products of the two Swiss. What is not shown are economic and social interdependencies or any other factors that keep the business economically afloat. Rather, they search for a necessary stage to formulate their biographical self-design in the unfiltered austerity and symbolically not yet fully used geography of the local. In their self-stylisation, the actors produce a social arena, a mental territory within which they relish re-coding the social hardships, deprivations and stagnation situations as a subtle stimulation potential.

The counter-horizons are the common myths of Berlin, the «city as an island» that is to be conquered, «Berlin mentalities», but also structures of local opportunities brought about by economic decline in the perforated – and declining – workers' boroughs of Berlin. Hence, the base motive correlates with Berlin metaphors of the early 1920s, during the Weimar Republic, when individual and also entrepreneurial self-assertion as both a cultural and an urban project, fully caught on for the first time, and fully unfolded its potential.

The high degree of adjustability between professional and biographic situations is combined with the geographical place-making of the 'Substrat' enterprise. The economic transformation context of Berlin, with its neo-liberal and extremely flexibilised labour market demands in times of scarcity is translated as an invitation to cultural self-assertion and entrepreneurial self-realisation. The territory that is not yet fully transformed and encoded is identified as a stimulating milieu, with references to the 1920s, when the city was made out as an adventure playground and site for new cultural projects.

CULTUREPRENEURS ? PERFORMING MICROGLOBALISATIONS!

Empirical field material made it possible to show for the first time that, for the city of Berlin, which has long since been an advertisement for the geography of various cultural niches, Culturepreneurs' individual emplacement strategies are not to be read as just the product of neo-liberal policies and processes. Attaining autonomy has been forced upon them by a need to secure a livelihood. Culturepreneurs embody a highly ambivalent relationship: the catchword «new entrepreneurship» demonstrates individualised marketing strategies and social hardships. It also indicates the temporarily skilful alternation between different modes of institutional integration (McRobbie, 1998, p. 14; McRobbie, 2002, p. 476).

The spatial practices of these urban pioneers provide insights into new urban ways of behaving, helpful to analyses of communal culture. They also allow, what Angela McRobbie (2004, p. 3 ff) named «cultural individualisation», the observation of the playful (self) production and performance tactics of these individuals on the urban stage. These tactics reveal consciously constructed identificatory opportunities for adoption and adaptation, deliberately littered with contradictions for the purpose of fine distinction. The spatial practices and entrepreneurial activities are treated as significant changes in the reconfiguration of the organisation of work, relative to space and place, and focus on how these subjects operate in often-precarious existential life situations.

The empirical material demonstrates that there is as yet no professional category for the «curator», «project manager», «artist», «website designer» who is transparently multi-skilled and ever willing to pick up new forms of expertise. Such an individual is also constantly finding new niches for work and thus inventing new jobs for him/herself (e.g. «incubator», «creative agent» etc.), s/he is highly mobile moving from one job or project to the next, and in the process also moving from one geographical site to the next. Social interaction is fast and fleeting. Friendships need to be put on hold, or suspended on trust, and, when such a non-category of multi-skilled persons is extended across a whole sector of young working people, there is a sharp sense of transience, impermanence, and even solitude (Augé, 1995).

The field material described characteristics and ways of perceiving through which *Culturepreneurs* make themselves known as a new urban type on the urban stage: they form a new type of relationship between their work practice, entrepreneurial turnover and their own social and creative development. This set of activities must – according to the authors observations – be framed by and tied to a tension-filled, ambivalent self-made ensemble of spatial images and codes, which is difficult to interpret from the outside but is, and this is crucial, interpretable for insiders.

Ronald Hitzler's sociological interpretations are blind to space though they do not take into account the spatial dimension when analysing symbolic differentiation processes (Hitzler, Bucher, Niederbacher, 2001, p. 26). On the one hand, according to the author's findings and the observed agents in Berlin, these processes run on the basis of the readability of the physical environment without which inclusion and exclusion on the social-symbolic level would not be possible. On the other hand, statements made by *Culturepreneurs* indicate a playful attitude towards these very codes, which are sometimes connected associatively and sometimes ironically instrumentalised in order to express their own placing strategies. Bourdieu's 'fields' and sorts of capital (cultural, social or economic) are reflectively «sampled» for purpose of individual emplacement strategy as well as specific project-oriented needs.

Culturepreneurs represent the spread of a model where the biography of work is derived from the kinds of lives led by classic artists. The job market for artists generally has long been one of the most dynamic and flexible part-time job markets, which has ever existed. Discontinuous careers are the rule here; frequent changes between employment and non-employment, and between varieties of forms of work is the order of the day (Wiedemeyer, Friedrich, 2002, p. 163). *Culturepreneurs* have adopted this model in all of its contrasting facets. Their masterly marketing of their own labour is set against an existential insecurity, hidden by a playful Bohemian attitude.

Expert productions within the economy of attention hide a struggle to maintain one's own position in society. As creative labour entrepreneurs, *Culturepreneurs* (are forced to) take on the role of forerunners in the flexibilisation of the job market, a flexibilisation, which will, in all probability, grow to encompass other sections of the service economy (Wiedemeyer, Friedrich, 2002, p. 167 ff).

This economic interface function of the *Culturepreneurs* is also clear at the dissolving borders between mainstream and subculture. Once, youth practices and subcultural practices served as a means to distinguish their practitioners from those in mainstream formations, but now maintaining this form of demarcation is, according to my findings, becoming ever more difficult. The old dichotomy has been replaced by new social formations which no longer display a rigid contrast between mainstream and subculture, and profit-oriented service provision and cultural production, and which – through the constant and simultaneous processes of reshuffling and recoding – mediate between different social groupings. Those on the left complain about the much cited «sellout of the underground», but it is also true that no clear mainstream can be recognised either. Difference, rather than adaptation, is the main drive behind post-modern consumption and this has led to a hybridisation of the mainstream and to a multiplicity of heterogeneous styles and groups. Hitzler and Pfadenhauer (opt., p. 90) speak of «post-traditional forms of community formation», in which post-modern concepts such as individuality and community combine to produce a loose temporary structure, which is only momentarily binding.

Their functioning can generate affective identification processes only retrospectively. The linguistic analogy between the sociological category «Szene» [German for *scene*] and the spatial category «in Szene setzen» [German for «*be the centre of interest*»] links scenes to the radiation of a physical place. Club events, gallery, shop, exhibition and office openings, for example, are stagings and temporary place-makings of scenes on the urban stage, where actors use the urban fabric, the city or concrete buildings, to create networked relationships of power, meaning and tension in order to test new products. This social formation «scene» experiences and performs itself in its materiality and corporeality through its emotional presence at and with the places it selects. Consciously constructed places enable individuals to see and to be seen. These protagonists are at once both, participant and spectator, both equipped with subtle knowledge and skills of knowing how to get «in the scene» and «staying out» of (other) scenes. According to Blum (2001, p. 18), «scenes evoke the sign of tribal hegemony because their practices always means the rule of a specialized solidarity at that site».

The *Culturepreneurs* studied in this research project take on a central role in the constitution of professional scenes, in particular through helping to develop new urban coding formations. The synthetic (pioneering) achievement of the *Culturepreneurs* lies in the fact that they stage new tension-loaded and ambivalent location images and motifs even in places that have fallen out of the traditional logic of urban use. Existing urban material is brought into relationship with their entrepreneurial as well as artistic activity and contributes, in combination with what is physically present, to an ambivalent visibility of the location. This (locational) policy of temporary hiding and disappearance must be interpreted in the context of the development of heterogeneous scene practices. Hitzler/Bucher/Niederbacher (2001) identify these practices for the most part a-physically and in unclear relation to built space and not just to social space. The findings of the *Culturepreneurs* presented here, using the example of Berlin, demonstrate, that they use their respective localities especially for entrepreneurial activities. First of all, they build up that tension-loaded relationship which guarantees them artistic and entrepreneurial attention in the wider Berlin scenes. In this sense, the protagonists are representatives of a «network sociality» (Lash, 2000; Wittel, 2001, p. 75). Relational global networking and novel creative place-making methods on the local level are their vehicles to overstep the classical «handiwork» concept and to redefine the relationship between art and economy, subculture and mainstream, city and city image, and city and individual.

The fact that greatly distorted, sometimes romantic, often very imaginative spatial images are thus designed, all of which flirt with the socio-political realities, cannot be attributed either to a hedonistic outflow from the fun society, or to the spatial blindness of the *Culturepreneurs*. It is rather the case, that these *Culturepreneurs* prove themselves to be the architects of spatial scavenging and recycling. As *space pioneers,* they position themselves in perforated places of the city, places which, through deindustrialisation and reorganisation of the infrastructure, have fallen out of the cycle of economic use and out of the everyday awareness of urban society. Apparently functionless spaces, useless, neglected, leftover, and forgotten places have come to exist here. In short, inner city micro-peripheries are thus reconstituted. In an age of ever more closely controlled, staged shopping paradises and Disneyfied city areas, the *Culturepreneurs* conjure up memories of the instabilities of the face of the 19th century city, by means of temporary use, locational politics of hiding, and spatial visions.

Culturepreneurs may be considered as social switchpoints in an individualised society, in which new formations will be tested, and scenes formed and opened. Their entrepreneurial activity is characterised by fast moving fluctuations in spatial location. The mechanisms driving this rapid change may be sought in the spatial potential, as well as in the relatively unclear future of the city of Berlin. They might also be found in the *Culturepreneurs* themselves, in the ways in which they express their social integration in the game with existing powers.

Mediated by their institutional function as bridgeheads in establishing specific scenes, they develop spatial placing practices and movement performativity patterns which stand in ambivalent relationship to specific Berlin space strategies. The observed geographies of the ephemeral, of the temporary and of the selective compete with the spatial practices of the reconstruction of historical patterns, which is, in some locations (in the District Mitte, Prenzlauer Berg and others), already clearly visible. It is, of course, also (a little) cool to be against the dominant strategy and just as 'in' – by means of a tactically clever, oppositional concept – to profit from this financially (e.g. the Design Fair in Berlin, the «Berlinmai» in May 2004). The extent to which the attempt at spatial demarcation may be observed through the individual *culturepreneur's* own playing of the rediscovered distinctive location is also the extent to which the element of spatial flirting with the significance of the location and the spatial relationships comes to the fore in the game.

From this perspective *Culturepreneurs* also show strategies arising from need and, above all, they show the seeds of new kinds of urban acting. They also demonstrate how globalisation processes affect the politics of place and space with erstwhile-unknown formations arising and materializing on the micro-level, performing microglobalisations!

Bastian Lange
Institute for Regional Development and Structural Planning (IRS),
Department of Knowledge Milieux and Settlement Structures (FA 3),
Erkner, Germany
(Lange@irs-net.de)

NOTES

[1] This term reflects the importance of knowledge- and information-based service providers within an urban post-Fordist service economy, which has increased over the past ten years (Zukin 1998). It is from those innovative and flexible economies that cities are drawing their hope for economic growth and symbolic image gains. In this context the so-called local cultural industries – expressions of an ever-growing urban cultural sector – are more and more becoming the focus of attention (Bassett/Griffiths/Smith 2002, Pratt 1997).

[2] One might think on the new cultural evaluation of spaces, formulated in the term "place-making-strategies" (Healey et al 2002), or in policies of urban renewal. Many approaches are based on "creativity", social and cultural capital in order to mobilize existing and new potentials by processes of recoding to design new planning strategies aimed at solving spatially relevant problems.

[3] Cf. www.idz.de/designszene (retrieved on 2004-01-08)

[4] It shall suffice to refer here to Funke/Schroer (1998, p. 219 ff.) who do not assess Hitzler's dictum of a necessary integration into new forms of communalisation to be optional and to be conditioned on a purely emotional or aesthetical plane. Both authors opine that "sovereignty in issues of lifestyle is not superfluous luxury but competence of import for survival" (op. cit., p. 225). Hence, the socially differentiating criterion is less an apparently freely selectable subjective stylisation than an ambivalent "non-compulsory constraint towards necessary stylisation" of the self (op. cit., p. 227).

[5] Bette (1999, p. 101 ff.) shows this with the example of skateboarders who create a temporary arena in urban space where they (entertainingly) display their bodies – via acrobatic jumps – to the urban community. See also Borden (2001) for a general overview of the phenomenon of the appearance of skateboarders and their practices in urban space.

[6] German for 'hood'.

[7] The empirical material for the following analysis are based on 25 interviews of participants chosen by virtue of their entrepreneurial and artistic activities, conducted between September 2002 and April 2003 in Berlin by the author.

[8] Spielfeld is a company and event agency communicating and managing cultural projects in socio-economic instable housing areas of Berlin. (retrieved on may 15th 2004, from www.spielfeld.net)

REFERENCES

Appadurai-Breckenridge, C., Bhabha, H. K., Chakrabarty D., Pollock, S. (eds.) (2002). *Cosmopolitanism*. Durham, NC: Duke University Press.

Augé, M. (1995). *Orte und nicht-Orte. Vorüberlegungen zu einer Ethnologie der Einsamkeit*. Frankfurt on Main: Campus Verlag. Translated by Michael Bischoff.

Banks, M., Lovatt, A., O'Connor, J., Raffo, C. (2000). Risk and trust in the cultural industries. *Geoforum* 31 (4), pp. 453-464.

Bassett, K., Griffiths R., Smith I.(2002). Cultural industries, cultural clusters and the city: the example of natural history film-making in Bristol. *Geoforum* 33 (2), pp. 165-177.

Baumann, Z. (1995). *Ansichten der Postmoderne*. Hamburg: Argument.

Beck, U. (2001). Living your own life in a runaway world: individualization, globalization and politics. *Archis*, 5, pp. 17-30.

Bette, K.-H. (1999). Die Rückeroberung des städtischen Raums, in Bollmann (ed.). *Kursbuch Stadt. Stadtleben und Stadtkultur an der Jahrtausendwende*. Köln: Bollmann Verlag, pp. 101-114.

Blum, A. (2001). Scenes. *Public*, special issue of *Public*, ed. by Janine Marchessault and Will Straw, pp. 7-36.

Böhme, G. (1995). *Atmosphären*. Frankfurt on Main: Suhrkamp-Verlag.

Borden, I. (2001). *Skateboarding, Space and the City: Architecture and the Body*. Oxford: Oxford University Press.

Bourdieu, P. (1986): The Forms of Capital, in Richardson (ed.). *Handbook of Theory and Research of the Sociology of Education*. Westport: Greenwood Press, pp. 241-258

Bude, H. (2001). *Generation Berlin*. Berlin: Merve Verlag.
Casey, C. (1995). *Work, Self and Society. After Industrialism*. London and New York: Routledge.
Certeau, M. de (1988). *Die Kunst des Handels*. Berlin: Merve Verlag.
Cook, I., Crouch, D., Naylor, S., Ryan, J. (eds.) (2000). *Cultural Turns/Geographical Turns. Perspectives on Cultural Geography*. Harlow: Prentice Hall.
Davies, A./Ford S., (1998). Art Capital. *Art Monthly*, (1), p. 213, pp. 12-20
Fischer, G. (1998). Editorial. *Econy*, (1), pp. 1-3.
Forman, R. T. T., Godron, M. (1986). *Landscape ecology*. New York: John Wiley & Sons.
Funke, H./Schroer M.,(1998). Lebensstilökonomie. Von der Balance zwischen objektivem Zwang und subjektiver Wahl, in Hillebrandt/Kneer/Kraemer (eds.): *Verlust der Sicherheit? Lebensstile zwischen Multioptionalität und Knappheit*, Opladen: Leske und Budrich, pp. 219-244.
Giddens, A. (1990). *Consequences of Modernity*. Cambridge: Cambridge University Press.
Grabher, G. (1998). Urbi et Orbi. Local Economies in Global Cities. *Der Öffentliche Sektor*, 24, (2/3), pp. 1-24.
Grabher, G. (2001). Ecologies of Creativity: the Village, the Group, and the Heterarchic Organisation of the British Advertising Industry. *Environment & Planning A,,* p. 33, pp. 351-374.
Grésillon, B. (2004). *Kulturmetropole Berlin*. Berlin: Berliner Wissenschafts-Verlag.
Haber, W. (2001). Kulturlandschaft zwischen Bild und Wirklichkeit, in Akademie für Raumentwicklung und Landesplanung (ed.): *Die Zukunft der Kulturlanschaft zwischen Verlust, Bewahrung und Gestaltung*, 6-29. (Forschungs- und Sitzungsberichte, ARL; Bd., p. 215)
Healey, P., Cars, G., Madanipour, A., de Magalhaes, C. (eds.) (2002). *Urban Governance, Institutional Capacity and Social Milieux*. Aldershot: Ashgate.
Helbrecht, I. (1998). The Creative Metropolis. Services, Symbols and Spaces. *International Journal of Architectural Theory*, 3 Vol. 1 (retrieved Mai 15, 2004, from http://www.tu-cottbus.de/BTU/Fak2/TheoArch/wolke/X-positionen/Helbrecht/helbrecht.html)
Hitzler, R., Pfadenhauer. M. (1998). "Let your body take control!" Zur ethnographischen Kulturanalyse der Techno-Szene, in Bohnsack/Marotzki (eds.): *Biographieforschung und Kulturanalyse*. Opladen: Leske und Budrich, pp. 75-92.
Hitzler, R., Bucher. T., Niederbacher, A. (2001). *Leben in Szenen. Formen jugendlicher Vergemeinschaftung heute*. Opladen: Leske und Budrich.
Holert, T. (ed.) (2000). *Imagineering. Visuelle Kultur und Politik der Sichtbarkeit*. Cologne: Oktagon Verlag.
Holert, T., Terkessidis, M. (eds.) (1996). *Mainstream der Minderheiten: Pop in der Kontrollgesellschaft*. Berlin: Edition ID-Verlag.
Internationales Design Zentrum (eds.) (2003). *Designszene Berlin*. Berlin. (retrieved April, 14, 2004, from www.idz.de/designszene)
Koppetsch, C., Burkart, G. (2002). Werbung und Unternehmensberatung als "Treuhänder" expressiver Werte? Talcott Parsons' Professionssoziologie und die neuen ökonomischen Kulturvermittler. *Berliner Journal für Soziologie*, 12 (4/02), pp. 531-549.
Krätke, S. (2002). *Medienstadt. Urbane Cluster und globale Zentren der Kulturproduktion*. Opladen: Leske und Budrich.
Lash, S. (2000). *Network Sociality*. (Unpublished Paper at the Goldsmiths College London). London.
Lash, S., Urry, J. (1990). *Economies of signs and symbols*. London: Routledge.
Landry, C. (2000). *The Creative City. A Toolkit for Urban Innovators*. Comedia: Earthscan Envirnomental Books.
Leadbeater, C. (1999). *Living on Thin Air*. London: Viking.
Löw, M. (2001). *Raumsoziologie*. Frankfurt am Main: Suhrkamp-Verlag.
Matthiesen, U. (ed.) (2002a). *An den Rändern der Hauptstadt*. Opladen: Leske und Budrich.
Matthiesen, U. (2002b). Zur Methodik sozialräumlicher Milieuanalysen – Anmerkungen zur Rekonstruktion von Fallstrukturen sowie zur Praxis der Typenbildung, in Deilmann, C. (ed.): *Zukunft – Wohngebiet: Entwicklungslinien für städtische Teilräume*. Berlin: VWF Verlag für Wissenschaft und Forschung, pp. 119-136.
Matthiesen, U. (2002c). Transformational Pathways and Institutional Capacity Building: The Case of the German-Polish Twin City Guben/Gubin, in Healey, P., Cars, G., Madanipour, A., de Magalhaes, C. (eds.). *Urban Governance Institutional Capacity and Social Milieux*. Aldershot: Ashgate, pp. 70-89.
Matthiesen, U. (2004). Das Ende der Illusionen - Regionale Entwicklung in Brandenburg und Konsequenzen für einen neuen Aufbruch, in SPD-Landesverband Brandenburg e.V. (eds.):

Perspektive 21. Entscheidung im Osten: Innovation oder Niedriglohn? Vol. 21/22. Potsdam: Weber Medien GmbH, pp. 97-113.

Matthiesen, U., Lange, B., Jähnke, P., Büttner, K. (2004). Zwischen Spardiktat und Exzellenzansprüchen. Wissenschaftsstadt Berlin. *DISP*, Special Issue (1) p. 156, pp. 75-87.

McRobbie, A. (1998). Kunst, Mode und Musik in der Kulturgesellschaft, in Hoffmann, J., von Osten, M. (eds.): *Das Phantom sucht seinen Mörder: Ein Reader zur Kulturalisierung der Ökonomie*. Berlin: B-Books, pp. 15-44.

McRobbie, A. (2002). Clubs to Companies: Notes on the Decline of Political Culture in Speeded Up Creative Worlds, in Bittner, R. (ed.): *Die Stadt als Event*. Frankfurt on Main: Campus Verlag, pp. 475-484.

McRobbie, A. (2004). *Creative London – creative Berlin. Notes on making a living in the new cultural economy.* (retrieved April, 5, 2004, from www.ateliereuropa.com)

Philo, C. (2000). More words, more worlds: Reflections on the "cultural turn" and human geography, in Cook, I., Crouch, D., Naylor, S., Ryan, J. (eds.) (2000): *Cultural Turns/Geographical Turns. Perspectives on Cultural Geography*, Harlow: Prentice Hall, pp. 26-53.

Pratt, A. (1997). *The Cultural Industries Sector: its definition and character from secondary sources on employment and trade, Britain 1984-91.* Research Paper in Environmental and Spatial Analysis, Nr. 41, London School of Economics. London.

Scott, A. (2000). *The Cultural Economy of Cities. Essays on the Geography of Image-Producing Industries*. London: Sage Publications.

Thrift, N. (2000). Introduction: Dead or alive?, in Cook, I., Crouch, D., Naylor, S., Ryan, J. (eds.) (2000): *Cultural Turns/ Geographical Turns. Perspectives on Cultural Geography*, pp. 1-6.

Verwoert, J. (2003). Unternehmer unserer Selbst, in Verwoert, J. (ed.): *Die Ich-Ressource. Zur Kultur der Selbst-Verwertung*. München: Volk Verlag, pp. 45-55.

Wiedemeyer, M., Friedrich, H. (2002). *Arbeitslosigkeit, ein Dauerproblem. Dimensionen, Ursachen, Strategien*. Opladen: Leske und Budrich.

Wigley, M. (1998). Die Architektur der Atmosphäre. *Daidalos* (68), pp. 18-27.

Wittel, A. (2001). Toward a Network Sociality. *Theory, Culture and Society*, 18, (6), pp. 51-77.

Zukin, S. (1998). Städte und die Ökonomie der Symbole, in Göschel, A., Kirchberg, V. (eds.): *Kultur in der Stadt. Stadtsoziologische Analysen zur Kultur*. Opladen: Leske + Budrich, pp. 27-40

DENIS COSGROVE

LOS ANGELES AND THE ITALIAN CITTÀ DIFFUSA: LANDSCAPES OF THE CULTURAL SPACE ECONOMY

Air View of East Los Angeles, source Veronica della Dora.

This essay examines two regions, which, while geographically and historically distinct, share morphological characteristics that make them significant locations for understanding landscape implications of the 'new cultural economy of space'. The metropolitan region in Southern California centered on Los Angeles and the Italian region of Veneto, lying between the Alpine foothills and the Po River delta inland

69

T.S. Terkenli & A-M. d'Hauteserre (eds.), Landscapes of a New Cultural Economy of Space, 69-91.

from Venice occupy a similar land surface: some 25,000 square kilometers. In their different ways, each has been represented as a paradigm 'post-modern' landscape, although they seem at first glance utterly distinct in historical evolution and scenery. The LA region has a recorded history of human settlement dating back only to the late 18th century, and the speed and nature of its urbanization over the past century have continuously challenged received ideas of what constitutes urban space and society. Faced with the dizzying glamour and onward rush of cultural innovation in Southern California, its specific cultural history is often neglected or forgotten. The Veneto, by contrast, was already colonized and urbanized in the days of the Roman republic. Its centuriated fields, feudal strongholds, patrician villas and the civic architecture of its venerable urban centers – Padua, Verona, Venice itself – have long figured the Veneto as a regional showcase of European heritage. The vibrancy of its contemporary space economy has led students of post-modern space to label this part of North-east Italy an 'exploded city' (*una città diffusa*), finding there a fascinating laboratory for exploring post-industrial landscapes in early 21st century Europe. The hypermodern aspects of the Veneto however, can be overlooked by visitors and students dazzled by its historical depth, or, if noted by such observers, only with concern and regret for a supposed authenticity of a lost landscape.

The value of placing Southern California next to the Venetian *città diffusa* derives from juxtaposing two superficially very different landscapes in order to reveal significant similarities in their geography and evolution that might suggest lessons in understanding and managing the new cultural economy of space. By examining the landscape history of the Los Angeles region, I am not suggesting that it represents an urban future for the Venice region, nor that the Veneto may be read as simply an earlier version of Southern California landscape. Rather, I suggest that social and cultural (as well as economic) processes often regarded as wholly contemporary may not be entirely so, and that their impact on these two regions has produced some important similarities in landscape outcomes, at least in coarse grain and viewed at the regional scale. These similarities, as well as some significant differences, help us better understand forces that are shaping the post-industrial landscape more generally as 'a complex and highly attractive mix of old and new, familiar and culturally different, real and staged, compressing and condensing geographically distinct versions of the world into single landscapes' (Terkenli, 2005, 1). While the forces of contemporary change inevitably generate tensions, to suggest that their only outcome is 'loss of place', 'unworldment', or 'inauthenticity' represents a failure of imagination and empathy with the richly varied ways that people live in and continuously manipulate their worlds, in favor of an intellectually fashionable Heideggerian angst.

I open with a description of the two regions as postmodern settlement landscapes, focusing on key similarities and differences. I then offer a brief survey of their shaping historical geographies, paying specific attention to key 'cultural visions' that influenced the regions' landscape images at key moments in their respective development, notably the vision of a leisured lifestyle in a rural, but not actively agrarian setting. I turn finally to some aspects of the management of these spaces paying special attention to their communications landscapes and spaces of domestic ecology.

LOS ANGELES AS A POSTMODERN LANDSCAPE

The erasure of territorial boundaries as containers of spatial forms and processes is a defining feature of the new cultural economy of space. Such absence is immediately apparent in the urbanized landscape of Southern California. The city of Los Angeles, home to some 3.5 million people occupies a bizarrely shaped section of Los Angeles County[1], whose 100 plus incorporated cities and municipalities are home to over ten million people drawn from every ethnic and national group on earth. This countywide conurbation in turn lies central to a 200km zone stretching east from the Pacific Ocean, through the 'Inland Empire' cities of Pomona and Ontario into the Coachella Valley settlements of Palm Springs and Indio, and north to south for 120 km from the northern Antelope Valley and Ventura County to the southern parts of Orange County. Some fourteen million people live, work and play in this region, a number increased weekly by over one thousand new immigrants.

To refer to this area as an urbanized zone however is not to imply a continuous built environment. The LA region is a complex jigsaw of densely settled spaces in the open floodplains and basins of Southern California, with remnant agricultural zones of commercial farming, and hill, mountain and hot desert environments up to 3000 m elevation, used for water catchment purposes, ecological conservation and recreation. Wilderness spaces, more isolated than many to be found around the Mediterranean, lie within less than an hour's drive of downtown LA. And to the south and west extends the Pacific Ocean, its islands, beaches and coastal waters, another critical environmental and recreational resource. As is also true of the Veneto, there is great complexity of administrative structures, so that while the City of Los Angeles is the dominant center within its region, a mosaic of overlapping administrative autonomies from federal to local scales counters the ability of any single authority to impose a unitary political will or set of solutions over the environments and spaces over the region. Rather, the activity space is held together by a sophisticated system of road, rail and electronic communications that offers an exceptionally high degree of mobility for its population's varied activities. That population benefits from levels of education and skill, personal affluence and consumption that further generate mobility while placing heavy demands on all aspects of the natural and human infrastructure: America's largest port at Long Beach-San Pedro combines gateway shipping for manufactures, oil terminal and refining capacity and tourist cruising. Two international airports (LAX and Ontario), three other national airports and a scatter of local fields are crucial to the regional economy and provide high levels of global and continental connectivity. Private and commercial access to advanced technology and communications media is greater and more widely diffused than in any comparable population concentration on earth, aided by a dense concentration of universities, laboratories and research centers, and the status of Hollywood and Anaheim as the global foci of the movie, TV and theme park industries – all defining components of the new cultural economy.

My emphasis on the cultural economy and communications is not intended to obscure the very important role of more conventional forms of economy in the LA region. The city is one of the most important manufacturing centers of the USA, the range of its products stretching from advanced space and military technology to sweat-labored garments and handicrafts. Whole municipalities are devoted exclusively to such activities: 'City of Industry' for example, whose residential population is a meager few hundreds. Nor should we ignore the continued role of agriculture in parts of the region: Ventura County remains the nation's principal lemon producing region, and citrus joins a cornucopia of fruits, vegetables and flowers to define even today tracts of garden landscape among Southern California's spreading bungalows and tract homes.

It is the cultural economy however that defines and gives shape to this landscape. Nor is this a recent shift away from older forms of land and life. The editors of this volume define the 'new' cultural economy of space in terms of new collective experiences and sense of place/time, *de-differentiation of public and private spheres* that rearticulate the relationships between the personal and the social, the *desegregation of the realms of leisure* from those of home and work, *disruption of conventional scales* of social and geographic relations by new modes of *communication and connection*, rapid and overarching exchange and communication of *symbolic goods* and the increased domination of *visual over textual media* (*Terkenli*: 2005, 2). By these criteria, the cultural economy of space has always characterized Los Angeles and the Southern California that emerged in parallel with it: from their late 19th century origins and throughout their astonishing 20th century growth into one of the world's greatest metropolitan regions. While the precise framing of the 'California Dream' has varied over time, Los Angeles has continuously been defined and developed in large measure to fulfil a distinctive set of cultural goals, within which a leisured lifestyle was by far the most important (McClung: 2000, Culver: 2004). Emblazoned in gold on the ceiling of Los Angeles City Hall, completed in 1931, is the motto: 'The city came into being to preserve life – it exists for the good life'. It is worth reviewing briefly the historical evolution of this vision of leisured life, seemingly unique half a century ago, today more and more obviously an early example of what has become a much broader phenomenon of post-modern space.

LEISURE IN THE MAKING OF SOUTHERN CALIFORNIA LANDSCAPE

The paradox of an intensely privatized land tenure system with severe limits to public ownership and thus to the emergence of public space on the one hand, and on the other an almost messianic vision of the moral community, captured by the phrase 'City on a Hill' is a characteristic of the United States generally. In California it is sharply and continuously apparent, from the irony of Hollywood's initial foundation as a temperance community, through the populism that flavors the state constitution, to the passage of Proposition 13 in 1978 restricting the ability of the state to raise money from property taxation, and the birth in California of the gated community and Homeowners Association. With Anglo annexation of this former province of Mexico and with statehood in 1849, the existing cadastral pattern of large Spanish land grants – *ranchos* – became the permissive framework within which the Anglo grid-iron land division and fee-simple purchase system was inscribed. Failure of commercial ranching in the 1870s coincided with the arrival of transcontinental rail links and the promotion of Southern California as an Edenic paradise for moderately prosperous Mid-Westerners seeking an escape from the rigors of continental winter, urban life in industrial St Louis, Chicago or Omaha, age and infirmity, especially tuberculosis. Up to a quarter of immigrants into Southern California before 1900 were invalids drawn by the supposed health benefits of its climate. Many others had first come as wealthy tourists attracted by the combination in America's 'Mediterranean' (or its 'tropics', or desert Holy Land, depending on which booster literature one reads) of winter warmth, spectacular scenery and a mythical 'Spanish' culture promoted by popular journalists such as Charles Loomis and retailed in romantic novels such as *Ramon* (Culver: 2004, DeLyser: 2003). Visitors were accommodated in fabulously grand hotels such as the Del Coronado in San Diego, the Glenwood Inn at Riverside, and the Raymond, Hotel Green and Huntington hotels in Pasadena.

Between 1880 and 1930, the population of Los Angeles County grew from 80,000 to more than two million. Before the dust-bowl migrations of the 1930s, a significant proportion of California's immigrants arrived with sufficient capital to purchase a small subdivision of land with crafted bungalow and citrus grove, located in one of the dozens of independent communities scattered across the LA basin, the San Fernando Valley, below the San Gabriel Mountains and into Orange County, with views of the surrounding mountains and a system of light electric tramways that could take them within the space of an hour or so to the city for shopping and entertainment or to the beach for recreation. By 1930 one in every three households there owned an automobile, principally as a leisure vehicle, for which newly built roads made the beaches, mountains and deserts easily accessible. If the Edenic vision of the leisured life in a land of endless sunshine was always matched by a darker side to Los Angeles, signified in virulent racism against Hispanic, Asian and Black Californians, high levels of violent crime and broken financial dreams, it remained powerful enough to generate styles of living that the glamour of 1930s Hollywood's screen images would project from Los Angeles across the United States and the world[2]. In Southern California the geographic nexus of land and life would be transmuted culturally into land*scape* and life*style* (Figure1).

Figure 1. Palm Springs, California (Chase Langford).

In communities with such evocative names as Arcadia, Naples, Florence, Venice and Alhambra, new modes of living shaped this distinctive landscape. California's sun, sea and scenery played a critical role, although the physical landscape was often marred in reality by oil wells sprouting among the citrus groves and wildfires in the suburban tracts. Yet design of the human landscape consistently sought to embrace and enhance Southern California's unique geography. Here, the single-story bungalow with its outdoor sleeping porch (a housing type designed for summer living in colonial India's hill stations) became a permanent family home. Here too, in the inter-war years, a series of historicist styles were standardized for the suburban home: the 'Spanish' with its white stuccoed arches and Roman tiles, the 'English' with its Tudor half timbers, the 'Hansel and Gretel' with its pseudo-Black Forest shingles. And in the Hollywood Hills and in post-war Palm Springs, architectural modernism created the horizontal 'ranch-style' of single storey, open-plan house whose picture windows were disposed in order to frame scenery, and blur the boundaries of interior and exterior space.

Of the liminal architectural spaces that fused interior and exterior the most iconic was the swimming pool. David Hockney's *Splash* with its turquoise pool, azure sky and lazy palms captures the essential elements of Los Angeles' leisured domestic image at mid-century. In 1949, there were some 10,000 private swimming pools in USA; by 1959 the figure was 250,000, of which 90,000 were located in the city of LA alone (Culver: 2004, 282). The casual dress codes that had first emerged around the pool and in the beach and desert leisure communities became pervasive in all parts of the suburbanized city, eroding performative distinctions between public and private space. This however merely reflected new forms of spatial organization in the built environment itself. As the original scatter of communities coalesced and residential subdivisions replaced the citrus, olive and avocado groves, so the characteristically 'post-modern' form of urbanism associated with Southern California came into being: low-rise, polycentric and non-hierarchical, stitched together by highways, lacking traditional forms of urban public space in the form of squares and monuments, pedestrian streets, parks and fountains, and good public transportation. The palm-tree is appropriately emblematic. Some 25,000 were planned along the boulevards of Los Angeles for the 1932 Olympic Games. They offer neither shade nor fruit, and their serried ranks belie the image of oasis or palm-fringed island with which they are normally associated. But in the commercialized city, they do not block commercial frontages for passing motorists and they offer an urbanized version of fantasy vacation space (McClung: 2000, 125).

In place of traditional elements of civic space, Los Angeles offered precisely this (at least to its mobile and affluent its citizens): a landscape of privatized leisure, of endless play in the golden sunsets: beyond the sprinklers playing across iridescent green lawns and exotic verdure, within natural spaces and environments carefully preserved for active recreation, and through an ever-expanding array of commercial leisure opportunities, many placing equal emphasis on the personalized culture of 'the body beautiful'.

Two examples suffice to indicate the way that Southern California landscapes, shaped by its peculiar attachment to leisure culture, have now become defining features of the new cultural economy of space across America and beyond. These are the golf-course suburb and the theme park. In the late 1940s, a group of actors and movie industry associates, attracted to Palm Springs in its heyday as a relaxed vacation spot within easy reach of Hollywood, purchased the Thunderbird Ranch for development as a country club, and hit upon a novel way of financing the venture. The golf course would be paid for by the sale of residential lots marked out and sold along the fairways and around the greens. House design was restricted to single story, low and rambling units, while a homeowners' association enforced deed restrictions governing the maintenance and appearance of private as well as 'public' space. The entire development was gated; the golf course, green with imported fescue and watered from deep desert wells, was the focus of its 'civic' life. Air conditioning and fast freeways to Los Angeles meant that the vacation home in the desert could become the permanent family residence. Partly through the televising of golf tournaments played at Palm Springs, the Thunderbird Ranch came to represent a leisured lifestyle option promoted and desired across America (Culver: 2004, 288-91). Its success as a symbolic landscape is reflected in Ford Motor Company's choice of 'Thunderbird' as the name for its 1955 sports car. By the late 1970s, the gated golf-course suburb had become an ubiquitous element of the postmodern landscape, at least around the Pacific Rim.

The theme park too appeared first in Southern California, in the same decade as Thunderbird Ranch. It would become a defining leisure space of the new cultural economy. In 1955, Walt Disney, who had made his fortune in animated cartoons, purchased a 180 acre site of orange and walnut groves in Anaheim, and drew upon Hollywood movie-lot design skills to create five permanent, stage-set 'landscapes':

'Main Street USA': a pastiche of small town America in the early 20th century, a tropical Afro-Asian 'Adventureland', 'Frontierland' based on the mythical American West, a 'Fantasyland' composite of children's storybook places, and 'Tomorrowland,' that gave geographic form to America's fascination with 'space-age' science and technology. Each of these combined sanitized and bowdlerized place simulacra with fairground rides, and performances and parades featuring cartoon and fairytale characters. Promoted through television, the enormous commercial success of Disneyland led to the rapid development of other theme parks under the sunny skies of Southern California: Universal Studios Hollywood, Seaworld at Long Beach, Knotts Berry Farm and Magic Mountain, all of them drawing heavily on a leisure-hungry metropolitan population as well as acting as a principal attraction for tourists from the USA and beyond. In the past quarter century, the theme park has evolved in close association with other motors of the new cultural economy, especially retail consumption, residential development and the movie industry into a defining element of post-modern space. Orlando in Florida has grown into a 21st century metropolis on the foundation of Disney resorts, which have 'imagineered' new modes of civic life and built environment: the 'new urbanism' (Katz: 1994). It is now possible to live permanently in the fantasy landscape of imagined small town America circa 1925. Los Angeles' newest retail mall, 'The Grove', completely blurs the distinctions between theme park architecture, heritage, and entertainment (rides, water displays and performances) and shopping spaces.

Southern California thus continues to pioneer the lifestyles and landscapes of the new cultural economy. To the historical reasons for its emergence as a new type of landscape: a climate and physical environment both exotic and appealing to European people, the promotion of leisure as a 'lifestyle option' to a white middle class with the resources and mobility to invest in such a 'dream', the existence of a (largely non-white) service class whose low wages could subsidize such a mode of living, and cheap, private transport heavily subsidized by a federally-financed highway infrastructure, have been added new ones that continue to generate cultural innovation in the region. Perhaps the most critical is the continued growth and diversity of its 21st century population. There is no longer a single majority ethnic or linguistic group in Los Angeles, or increasingly in any part of the metropolitan region. A 'cauldron' of diverse ethnic, linguistic and religious groups is further segmented by openly celebrated diversity in other aspects of culture: sexuality, diet, fashion and lifestyle identities (Clark: 1998). Segmentation, fusion and hybridity all occur simultaneously, leading to continuous cultural innovation. At the same time, the appeal of certain long-standing elements of Californian living: single family residence, auto ownership and use, outdoor recreation, and body culture seem to hold a quasi-universal appeal among the region's residents, regardless of their cultural origins. Liberal social values and fiscal conservatism also remain deeply rooted.

Southern California's many critics claim that the outcome of these values and their landscape impacts is banal at best, and at worst corrosive of the distinctiveness and 'authenticity' associated with traditional and more stable places and landscapes. Here, for them, is the epicenter of enworldment and thus of the processes of unworldment and their collapse in turn into landscapes of deworldment: 'fictitious, commercialized, disposable, ephemeral, staged and inauthentic.' The force of this criticism might perhaps best be judged by comparing Southern California to the *città diffusa* of the Veneto.

La città diffusa – a post-modern landscape in Italy's Venice region

The physical geography of the Veneto bears some superficial comparison to that of Southern California. A wide lowland basin interrupted by piedmont spurs of the Alps and isolated hill massifs, and divided into zones of dry gravel outwash plains, reclaimed marshlands and coastal lagoons stretches some 150km east to west by 100km north to south. It is bounded to the north and east by the Alps whose white peaks are as intermittently visible from the hazy Venetian plain as those of the San Gabriel Mountains are from Los Angeles. To the east and south is the Adriatic Sea whose waters support a line of 20th century marine resorts similar to those that stretch from Malibu to Dana Point in California. While winters at this latitude are significantly harsher than those of Southern California, summers are similarly hot, if more humid, and both areas have developed through intensive intervention and management of the regional hydrology.

Aspects of the Veneto's contemporary human geography also suggest comparison with Southern California. The Veneto's population of 4.4 million people, while significantly less that that of Los Angeles county, is similarly distributed across multiple, formerly independent, urban centers and 580 *communi* that have coalesced economically into a polynucleated mosaic of increasingly interdependent residential, industrial, commercial, agricultural and recreational land use zones. In the three provinces of Venezia, Treviso and Padova are located 450,000 businesses, 97% of them with fewer than 15 employees (Erbani: 2002). They occupy a dense landscape of historic buildings, high-tech industrial and service units, commercial and family-based agricultural structures, modern residences and natural environments – mountains, hill slopes and wetlands – of high ecological and recreational value. Physical coalescence is producing a loose pattern of *rurbanization* strung along a non-centric network of roads, rail and information links. Thus on lands between Vicenza and Padua or Treviso and Pordenone that were purely agrarian until the 1980s, corn grows in a field adjacent to a smoked-glass superconductor factory, a gold jewelry workshop and a furniture hypermarket. The 16th century villa and its garden or the 18th century *barchese* stands next to a modern residence built on an artificial rise above the drainage ditches of a former rice field. Superhighway (*autostrada*) connections are few, their development limited by physical and legal constraints of a long-settled agrarian cadaster. Rather than channeling the urbanization of former rural spaces, modern infrastructure has followed a growth pattern largely determined by individual decisions in a region dominated by small property owners who have used their agricultural parcels for family industrial and commercial enterprises of the Italian model of domestic entrepreneurialism and self-employment (Dal Pozzo: 2002). Nevertheless, the decentralized, diffused city, through its varied morphologies, is characterized as an 'auto-habitat', its territory structured basically by individual capillaries of accessibility.

While domestic and local ties remain strong and residential mobility more limited than in the USA, the Italian region enjoys elevated levels of connectivity into a globalized world. The Veneto, like the LA metropolitan area, has a highly skilled population with elevated consumption values (per capita income at E23,000 is nearly 20% higher than the Italian average), high levels of auto ownership and ready access to the flexibility offered by information technology. It is one of Europe's most economically dynamic and materially successful zones, with a regional GDP that is among the highest in the EU (Erbani: 2002). An icon of its contemporary global success is the clothing company Benetton at Treviso, pioneer in flexible, computer-based design, just-in-time manufacturing and globalized retailing (Figure 2). Increasingly the Veneto offers the opportunity space of a single metropolitan area. Its citizens operate within this space as if it were a unitary city – living in Vicenza, working in Padua, with a second home in the foothills of Asolo, and enjoying the cultural life of Venice. The Veneto is not unique in these respects, it is one of the more economically successful of a number of such regions that have emerged within the new Europe over the past three decades, both within Italy (Tuscany, Emilia-Romagna) and beyond (Randstaad in Holland; the Rhur in Germany, and 'Transpennine' between Liverpool and Hull in northern England). Parallel to the de-territorialization of the pre-existing European geography of distinct urban and rural spaces within the relatively impermeable bounds of nation states is a process of re-territorialization along transport corridors and around constellations of interdependent centers of activity. EU planners and policy makers increasingly recognize these as significant spaces within Europe's evolving geography.

With economic success has come very rapid cultural and ecological change that has impacted directly on the landscape. Most dramatic is the erosion of the historical division between city and country – marked in a tradition of walled fortification that stretches from the crenellated medieval battlements of Marostica, Cittadella or Montagnana to the sophisticated geometries of bastions, redoubts and earthworks that circle Padua, Verona or Palmanova. Building simply sprawls loosely along roads and former farm paths. 32.9 million cubic feet of new residential and industrial construction occurred in 2002 alone (*Thiene.it*: 12/5/2004). Equally dramatic is the changing ecology of land, water and vegetation, as drainage ditches are covered and fields amalgamated, and new domestic plantings replace the cultivation of productive crops. To grasp the landscape impact of these changes, I offer a brief outline of the historical vision of landscape in the Veneto.

Figure 2. Old and new in the diffused city: Benetton store at home in the historic center of Treviso (Author).

PALLADIAN LANDSCAPE AND LEISURE

If a consumerist vision of leisured life played a vital role in shaping the historical evolution of the Southern California landscape, a more conventional and elitist, but equally powerful, vision of leisure helped shape the historical landscape of the Veneto. It continues to play a vital role in the region's success as one of Italy's most important recreational spaces, drawing over 8.7 million tourists each year, over 5 million of them from outside Italy, by far the largest number of any Italian region. The ultra-modernity of much of the 21st century Veneto landscape appears to threaten a heritage space that in previous work I have termed the Palladian landscape (Cosgrove: 1993). By Palladian *landscape* I mean more than simply the buildings and artistic monuments constituted by the patrician villas and estates that the 16th century architect and his imitators have left in locations scattered throughout the Venice region. Today, these are, with a few exceptions, restored and maintained as heritage sites, whether by public agencies or private owners. By Palladian landscape I mean to locate those distinctive architectural and design elements within a complex cultural geography that developed across the region during the 16th century and beyond, at the height of Venetian economic, political and cultural influence and which long thereafter dominated the shaping of both the material landscape and its representations.

The main features of this Palladian cultural landscape include the physical combination of *altopiano*, piedmont slopes and hill outliers, watered plains, marshlands and tidal lagoons already discussed and their exploitation for commercial agriculture by urban capital through large-scale land appropriation (via debt peonage) and estate consolidation, new forms of labor contract, private irrigation and large-scale, consortium drainage schemes, and the introduction of new crops such as rice and American corn. Venice, a dominant commercial city with global trading and geopolitical concerns in the Mediterranean and beyond, became increasingly tied to its land empire on the *terraferma*. Military defense, environmental protection (for lagoon and river regulation, and for preservation of naval timber), together with transalpine trade and securing the urban food supply, bound ever more closely the interests of the city and its hinterland. A dense network of formerly autonomous urban centers (Padua, Vicenza, Verona, Brescia etc) with strong historical identities and a high degree of administrative autonomy encouraged competition with both the *dominante* and each other, economically and for cultural capital. Between the mid-16th and the late 18th centuries, modernization of land measurement, classification and use planning through scientific survey and mapping found a parallel aesthetic expression in a widespread taste for landscape art in poetry and painting and villa culture (*villeggiatura*). This would come to influence elite landscape taste across Europe.

The architectural set pieces of villas, palaces and gardens and a tradition of landscape art were financed through the commercialization of the Veneto landscape. They represented a new anti-urban 'culture of nature', articulated through an ideology of 'holy agriculture' wherein the hard labor of taming nature and making it productive was undertaken by field workers (*braccianti*) carefully hidden from the patrician gaze across smiling fields of grain, blossoming orchards or ripening vineyards. That vision found expression in the architectural forms of the Palladian villa, in the landscape paintings of Giorgione, Titian, Paolo Veronese and a host of minor painters, and in the literature and practices of villa living. More than a purely local phenomenon, villa culture was nurtured in the context of an emerging globalization of European life and culture, apparent not only in geographical discovery and international trade, but in print publishing, religious reformation, humanism and the new science, to which key actors in both Venetian and provincial urban elites contributed significantly. In its rejection of the city and its embrace of a leisured vision of what the English would come to call 'country life,' this culture offered a prototype of the suburban and exurban family home and garden that would become a defining landscape element of the cultural economy of space (Dal Pozzolo: 2002, Bertuglia: 2002).

Today, this Palladian landscape is difficult for the casual observer to detect. It is obscured by a palimpsest of erasures and rewritings of the Veneto made during the succeeding 400 years. The pace of transformation has vastly accelerated over the past four decades – until the mid-20th century the main features of the agrarian landscape inherited from the Venetian modernization remained visibly apparent. Paradoxically, as the working agrarian landscape has been swamped by post-industrial sprawl, much of its signature architecture has been carefully restored. Initially this was limited to patrician villas and large *barchesse*, but increasingly, 0more utilitarian farm buildings are being recovered and put to modern, non-agrarian uses. To recognize other landscape elements however, such as property boundaries, hydrographic and cropping patterns, water technologies and such, demands a trained understanding of landscape morphology. Both historical study and an imaginative eye are needed to detect the original Palladian landscape, to hear the streams that irrigated a *brollo* below the windows of the Villa Barbaro, or view the open fields of irrigated rice stretching beyond, to witness the sawmill and paper manufactory in the Astico valley below the Villa Godi at Lugo, or to relive Hieronimo Godi's drunken and murderous fury over the rival Piovene family's claims on those mills.

However, recognizing the Palladian architectural heritage within a broader conception and scale of landscape allows us to understand the contribution that this historical intervention within the Veneto has made to the contemporary *città diffusa*. The Palladian landscape was a product of Venetian expansion into the *terrafirma* during an extended period of security and relative prosperity between 1530 and 1620. This established a tradition of political, economic and cultural coordination between formerly independent and competitive provinces. Thus Venetian authorities regulated trade and markets in the interests of the mercantile city, coordinated the construction of urban defenses and supervised an effective hydrographic planning system across the *terraferma* to protect the lagoon and secure the provisioning of Venice with cereals. Such economic and environmental co-ordination, continued by later Austrian and Italian states, provided early lineaments of the region's contemporary organization as a coherent space.

Material evidence of Venetian control is apparent in the *centri storici* of cities and small towns across the region: in symbols such as St Marks Lion in their *piazze*, in the adoption of architectural forms such as Venetian Gothic for their *palazzi*, in the fortifications of cities such as Padua, Treviso and Verona. These are now part of a common material heritage of the *città diffusa*. The early modern period also saw some coordination of communications across the Venetian plain, especially by water, through the establishment of the *Magistratura alle Acque*. Venetian investment in land on the *terraferma* introduced planned land management and improved access to the city along canalized streams on the mainland river system. Linkages between *terraferma* cities and the external world were increased by their immediate access to Europe's major trading city. Yet the Venetian pattern of administration left local cities a high degree of autonomy, their statutes and social hierarchies largely unaltered and in so doing engendered a structure of political complexity. While this legacy produces disadvantages today for large-scale, coordinated planning (road infrastructure, etc), it has equally significant advantages in permitting a flexibility for local initiatives that has been critical in the economic success of small businesses within the *città diffusa*.

The patrician villa, which lies at the heart of the Palladian landscape was a phenomenon of modernization. In addition to its residential functions it acted as a center of production. Architectural historians have commented on Palladio's success in combining the lagunar *delizia* and the *belvedere* style of the Classical villa with the mundane functions of the agricultural estate center. The resulting production space, set among highly managed commercial fields and serviced by a combination of small tenant farmers and day laborers was a focus of innovation, agricultural management techniques and artisanal skills. It could evolve into a broader proto-industrialization located in the villa landscape itself, as at Piazzola sul Brenta, or in the string of woollen, paper and leather mills between Schio and Thiene. The intimacy between agriculture and industrial production that characterized such structures remains visible today in the pattern of agro-industrialization in the Veneto's *città diffusa*. But while modernization and production were controlling aspects of the Palladian landscape, a vision of leisure dictated key elements of siting and decoration. For the patrician family the villa was a summer residence and escape from the administrative and environmental pressures of the city, and much of the culture of *villegiatura* revolved around such pastimes as hunting, reading, dining and conversation. There is a direct lineage between early modern Venetian vision of rural life as one of blossom-scented ease in the hill slopes of the Veneto and the Edenic dreams of early 20th Mid-Westerners purchasing smallholdings in Arcadia and Pasadena below the San Gabriel Mountains.

It is clear that both the environmental and cultural qualities that make the 21st century Veneto such a dynamic and economically successful region owe a major debt to the coordinating efforts of the Venetian empire in the prosperous 'long' 16th century. This is not to discount the significance of other historical interventions (the competitive autonomy of medieval *communi*; the administrative coordination of Austrian government; 20th century land reform and the light regulatory hand of the Italian State). But the historical significance of the 'Palladian' period lies in its powerful *landscape* expression of so many of the elements of unity in diversity that characterise the contemporary Venetian *città diffusa* and historically 'ground' its apparent ultra-modernity. This includes a remarkably rich cultural heritage of art and architecture, which remains a significant element of environmental quality, a major tourist attraction and a foundation of enduring local cultural identities.

UNDERSTANDING AND MANAGING DIFFUSED URBANIZATION

The landscapes of both Southern California and the Veneto have thus both been shaped in some measure by cultural visions originating in urban societies that saw individual land holding and a leisured rural – but not agrarian – life as desirable goals. Translating that vision into material form has produced many of the landscape features that render these regions highly attractive to a post-industrial economy whose production spaces are extremely footloose and locationally flexible, and which depend upon a mobile, educated and discerning workforce. Both regions present that 'complex and highly attractive mix of old and new, familiar and culturally different, real and staged' aspect of the cultural economy of space (Terkenli: 2005). Knowledge-based economic activity (information technology, biotechnology etc) is increasingly attracted to such 'quality of life' locations. Quality of life is notoriously difficult to define with precision, but in this context it seems to include climate, scenic diversity and environmental quality as physical attributes, and multiple leisure and recreational opportunities together with the cultural attributes of easy access to both sophisticated urban resources. Both the regions under consideration compete very successfully in such a context, not because of recent moves to exploit latent advantages but through long-standing aspects of their physical environments and evolution as cultural landscapes.

History, or more specifically 'heritage' – commodified and materialized history – is a significant element in the new cultural economy of space. The capacity of a region to exploit its past in ways that appear meaningful and satisfying to contemporary residents, as well as to investors, immigrants and tourists is widely recognized as a major competitive advantage for places and regions seeking to attract highly mobile capital.   Inevitably the public presentation of history as heritage is highly selective, emphasizing 'positive', romantic and heroic aspects and ignoring or repressing darker features of the past. Here too, we need to be cautious of assuming this to be a recent phenomenon. Within two decades of 1850 and the Anglo destruction of the Mexican economy and culture that had developed around the *pueblos* and *missions* of Southern California, a romanticized version of an arcadian 'Spanish' California was being used as a promotional device for attracting Anglo settlers (DeLyser, 2003). By the same token, Venice's architectural, graphic and performative self-presentation has always been shaped by consciously manipulated historical myth and heritage (Brown:1996). The very idea of rural arcadia that shaped the transformation of both the rural Veneto and Anglo settlement in Southern California has a lineage stretching back to Virgil and Pliny.

Given the historical depth of both the processes and forms of the cultural economy of space in Southern California and the Veneto some of the social assumptions adopted by critical commentators about 'unworldment,' the loss of deep attachments between local populations and their material landscapes, often signified by the diminution of the public realm, may be misguided. Geographers and other students of space are increasingly aware that the local is always global in the sense that it is never immune from processes and interactions that operate at greater scales that locality, and that both place and landscape are best understood as processes that, however discontinuous, are contingent and deeply historical. This is certainly true of both the regions under consideration. It does not of course mean that there are not always critical questions of managing change and seeking to understand and ameliorate its socially and geographically uneven effects. I consider briefly two of these that bear directly on the visible landscape: communications and domestic ecology.

COMMUNICATIONS

One key to the evolution and management of diffused urbanization within the contemporary cultural economy of space is communication: the creation of a surface which maximizes individualized movement, accessibility and communication, both spatial and social. Of course, Venice itself is often cited as the paradigm walking city with an historical tradition of balanced government whose primary goal was social stability. Its dense construction of residential and commercial properties focused on parish *campi* was connected by minor canals linked to the Grand Canal – the 'freeway' of the medieval city. The republic was organized as a highly managed collective social space with both its merchant nobility, its citizens and the common people of its parishes and *sestieri* regularly required to engage competitively in a public life of organized ritual, centered on the vast public spaces of San Marco. But, in the context of early modern urbanization, Venice was successful precisely because of the density and efficiency of its physical and social communications, both internally and externally. The ease of moving around the city derived in part from the capillary structure of canals and waterways and the protection offered by the lagoon, which removed the need for walls and internal barriers to mobility. And the very sophistication of Venice's social codes and rituals indicates the powerful presence of individual, entrepreneurial initiative in a city of quite exceptional ethnic, religious and linguistic diversity that had to be to be regulated for the goal of civic harmony.

A capillary-like, non-hierarchical communications structure also characterizes the modern diffused city. The Los Angeles Basin evolved as a mobility surface principally through the combination of an initial system of electric railways connecting a scatter of agricultural settlement settlements. While Los Angeles acted as a network center, its dominance was never total and the mobility and connectivity permitted by the light rail system encouraged speculative development housing in environmentally favored zones (Hollywood Hills, San Gabriel Mountains, coastal beaches). The automobile soon permitted individual movement beyond the station stops on the system and would eventually see the latter replaced by an equally non-hierarchical network of freeways. The original scattered pattern was emphasized by the early removal of large land blocks from residential development by oil drilling and movie studios.

By the 1940s, the scatter of settlements, the need to connect valley basins across hills and mountain passes, and the high levels of automobile ownership among a highly affluent population, as well as the interests of petroleum companies drilling cheap oil combined to promote a system of freeways to connect the region as a single communications surface. Freeway development was only one of a number of mid-20th century attempts to control the evolution of a diffused city by imposing a single plan, but it was virtually the only one to succeed. Its consequence was actually to *increase* the spatial extent of the urbanized landscape. While the freeway system originally envisaged was never fully realized, it was the element that permitted a single rurbanized space to emerge (Figure 3). Even today, the freeway system of the Los Angeles region allows a higher degree of connectivity than any other comparably sized urbanized area. It is the unifying element that cuts across the patchwork of political, administrative, functional and environmental boundaries that characterize the diffused city.

Figure 3. Construction of the Hollywood Freeway, Los Angeles, California (Spence Photo Archives, UCLA).

And this is true not only in terms of the freeway as a functional communications link. The freeway interchange has famously become a paradoxical symbol of the collective, public culture of this new type of city. Not only are Southern Californians famous for the ways that they conduct their social life on the freeway (phoning, preening, flirting) but the freeway represents the principal element of public landscape in the sense of a regulated civic space. There are no tolls, there is a ban on all forms of advertizing, high levels of design and planting and careful monitoring of air quality characterize their regulation. The landscape of movement has become the public space of the postmodern diffused city.

The Veneto today faces acute problems in catering to and managing the auto culture so intimately tied to the evolution of its diffused urbanization. Here, the historical depth of the humanized landscape has a direct impact that prevents the evolution of the kind of freeway system that acts as the life support system for urban Southern California. The density of individual property ownership, the size of land parcels, the complexity of administrative geography, and the scatter of urban elements across a non-hierarchical system of roads dating back to Roman centuriation make it prohibitively expensive to build such a road system. This, much more than some mystical attachment to place and rootedness accounts for the failure to accommodate car ownership and use with new highway construction.

Cultural features such as the tradition of returning home for lunch and siesta exacerbate a sclerotic road system. On the other hand, Los Angeles is famous for its freeway gridlocks, so that it is becoming increasingly apparent in both regions that road transport systems maintain a threshold level of congestion controlled more by aggregate journey times than by the theoretical capacity and efficiency of the system, and that further evolution of the communications surface is achieved through other modes such as mobile telephones and information networks.

DOMESTIC ECOLOGIES

The two regions face similar environmental challenges in protecting fragile ecologies on deforested mountain and hill slopes, regulated waterways and wetlands, and coastal lagoons and marine spaces from negative human impacts. These include pressures for construction, drainage and over-extraction of aquifers and pollution of water flows, petrochemical smog and degradation of former plant and animal ecologies. Here, I concentrate on a minor, but highly visible aspect: the changing selection of flora that comes with urbanization of agricultural land.

Rurbanization on individual smallholdings has transformed the landscape ecology of the two regions in parallel ways. For Anglo-settlers in Southern California, agricultural production was rarely the primary economic concern; the state's agricultural wealth has largely been generated on the irrigated fields of the Central Valley, and by large land corporations employing casual and ill-paid labor. The citrus, olive or nut plantations of Southern California were commonly subdivided within a generation or two for residential development, leaving a small family remnant. Orchards would be replaced by irrigated suburban landscape planting designed for aesthetic pleasure alone, thus emphasizing foliage and flowers. While native species such as live oak can produce highly attractive scenery on grazed hill and mountain land, and both desert and semi-desert species can bloom spectacularly after the winter rainy season, neither fulfils the garden aesthetic associated with Southern California's imagined Mediterranean and tropical geographies. Thus exotics have been imported from across the world: jacarandas from Brazil, pampas grasses from Argentina, bird of paradise and flowering shrubs from South Africa, gum trees (*Eucalyptus*) from Australia and various palms from Polynesia. The choice is eclectic and edenic, but with a strong flavor of an imagined European arcadia in the incongruous ubiquity of the fescue lawn. Attempts on eco-nativist or water-saving grounds to encourage garden design based on 'native' flora have met with very limited success.

In the Veneto, a similar process is apparent, much to the consternation of localists committed to regional biodiversity and biotic heritage. Fast-growing evergreen shrubs (*Lelandiae*) and palms are increasingly used to define the garden spaces of residential villas from which vines and their support trees have been removed. Around many of the new industrial and commercial structures, a corporate 'blandscape' based on a limited selection of globalized exotics is replacing poplars, willow, aspen and other trees that secured the banks of the drainage channels. And the ancient olive trees of Puglia are being uprooted and transported in large numbers for sale at hundreds of euros apiece to be planted in the unaccommodating soils of the Venetian plain. The aesthetic is in some respects different from that of Southern California, but it displays some fundamental similarities. In both cases, a garden ecology is a necessary complement to the pattern of individual domestic buildings set on small private land holdings. The principal product of the space is leisure and thus referencing is to an imagined geography, of elsewhere in either time or space.

This does not imply a lack of attachment to place, indeed such attachment is intensified as individuals experience the freedom and personal pride that comes from the ability to create personalized landscapes, a freedom previously available only to the wealthiest patricians (who themselves of course never hesitated to import exotics to realize their imagined geographies).

Many other features of the cultural economy of space can be highlighted and compared in these two regions. The extraordinary intensity of cultural activity in the past century within the Los Angeles area – including the movie and entertainment industries, private architectural initiatives of extraordinary creativity, the concentration of great libraries, art collections and universities – means that questions of cultural heritage conservation and consumption now demand increasing political energy in Southern California, perhaps a surprising shared question for the two regions. In physical landscape, the delicate balance of land and water along the coastal fringe, perhaps the most sensitive measure of the social and environmental health, remains a central management issue in both the Venice and Los Angeles regions.

CONCLUSION

The spaces of the contemporary cultural economy are highly complex, in both their functioning and their morphology. And there are certainly important differences between them, for example the zoned separation of industry and residence in Los Angeles is not reflected in the more promiscuous mixing of land uses in the Veneto, and Italy has yet to experience the levels of immigration long familiar in Southern California and thus remains ethnically and culturally more homogeneous. But both regions have been subject to critical appraisal on the basis of social, architectural and landscape concepts developed in the context of Modernity (Davis: 1992; Soja: 1996; Ludoviana: 1990; Detragiache: 2003). Most relevant in this context is the critique of placelessness as a characteristic social and landscape feature of the post-modern landscape (Relph: 1979; Augé: 1995). The ecological model that emerged alongside Modernity in the late 19th century served brilliantly to describe the latter's paradox of progressive spatial change resulting from capitalism's competitively creative destruction and nostalgic attachment to fixed 'niches' wherein life could be rooted in familiar place. Many of the neologisms generated to capture the processes and forms of contemporary landscapes: 'inauthentic place', 'non-place,' 'heterotopia,' 'hyperspace' implicitly assume the existence in the past of a more fixed and stable relationship between land and life, place and culture, locality and home. The result is a declensionist reading of 21st century landscapes of the 'new cultural economy.'

The comparison I have sought to draw between Southern California and the Venice region might give pause to such readings. Not only can we trace the roots of 'postmodern' space in Los Angeles deep into the (admittedly brief) settlement history of Southern California, but the defining vision of the leisured life and its characteristic consumption requisites (private land, single family home, mobility, scenery and environmental quality) was already clearly articulated in early modern Venice. The key differences are two: the social breadth of access to such positional goods, and the changing technologies of communication and transportation. In 1550, the choices of leisured life in a rural setting and of an environing architecture and garden landscape whose visual references were geographically and historically exotic, was the privilege of a small number of patrician families. In early 20th century California those choices had expanded to a large middle-class population. Today they are available to very large numbers of families across North America, Europe, Australia, significant parts of Asia and scattered locations elsewhere on the globe. And, in every case, the chosen landscapes of leisured living display common characteristics (as well as important distinctions). Below the veneer of similarity, however, there seems little evidence that people feel alienated or existentially displaced; rather, they seem to embrace the experience of privatised place with the intensity displayed in the past by the privileged few who could afford such ownership of place and landscape. Perhaps their principal failing is that of former owners and guardians of landscape: blissful and possibly culpable ignorance of the human and environmental cost of sustaining the leisured landscape, a cost from which modern technologies allow ever greater distanciation.

Neither Southern California nor the Veneto is thus unique as a landscape expression of the cultural economy of space. More and more regions today share their defining features. But they do capture very effectively aspects of a characteristic post-modern geography and some of the landscape questions they raise.

CODA – VENICE BEACH, CALIFORNIA

Venice CA is located on the Pacific shore 15 km west of downtown LA. Like Naples to the south and Florence to the east, it is indistinguishable today from the urbanized basin of Los Angeles. It was developed in the opening years of the 20th century in an environment of lagoon, salt marsh and sand spit, the speculative initiative of Abbott Kinney, a real-estate entrepreneur typical of his time and place, who saw an opportunity to make money from beachfront housing for a leisure culture. Faced with competition from such places as Santa Monica and Manhattan Beach, he needed a novel selling point. His choice was typical of what in today's cultural economy of space is called 'imagineering': the creation of a geographical simulacrum, in Kinney's case a new Venice along the Pacific shore, complete with canals, bridges and gondolas, Piazza di San Marco and Doge's Palace (Figure 4). From its inception, the place promoted a relaxed and tolerant lifestyle, appropriate to the romantic, fantasy Venice upon which Kinney based his themed community.

Figure 4. 'St Marks' – Venice Beach, California (Author).

Although Venice Beach attracted visitors to its pier and amusement arcades in the early decades of the 20th century, sadly for his vision of a permanent community along the canals, Kinney was unaware of two key environmental challenges: the absence of the tidal flow which cleans the lagoons of Venice proper, which left stagnant canals, unpleasant odors and mosquitoes, and the discovery of a rich oil field immediately below his land. Within a few years, his beachside homes looked out over oil derricks stretching from the shores into the coastal waters. Despite imaginative attempts to attract visitors and residents to the canals, by 1960 Venice CA was a seedy, impoverished and marginal neighborhood in the diffused city: a place of gang violence, drugs and poverty. The cast iron columns and capitals of its Doge's Palace were rusting below the weeds that grew from its salt-caked brickwork. More Malomocco than Riva degli Schiavoni, Venice CA was a dead corner of the vast, diffused city.

But this story has a happier ending – oil drilling came to an end after mid-century and its environmental impacts were cleaned up in the 1970s, while modern pest control keeps the canals free of mosquitoes. Starting as a center for 'alternative' lifestyle in the late 1960s, the community has gentrified rapidly since. Venice canal properties are today among the most expensive in Los Angeles, comparable to those in the original Venice, while the decaying Doge's Palace and the 'muscle beach' are among Los Angeles' best known cultural heritage and tourist sites. Like the Lido of Venice Italy, Venice CA has become a distinct place, occupying a unique niche within a cultural economy of space that stretch across over the urbanized surface from its canals, lagoons and decaying gothic capitals to the snow peaked mountains and the deserts beyond.

The distance between the two Venices as sites of cultural and environmental heritage and locations within the contemporary cultural economy of space may not be as great as we often like to think.

Department of Geography, UCLA, U.S.A.

NOTES

[1] The territorial shape of the city, which balloons from its downtown into the large but non-contiguous space in the San Fernando valley to the north, and stretches along a four mile long but few hundred meter wide neck south to the port at San Pedro, and is otherwise punctured by numerous independent municipalities and cities, is the result of a long a fractious politics of incorporation, whose outcome was often governed by control of water. See Fulton: 2001.
[2] On the 'noir' side of California landscape visions see, among others, Mitchell: 1996; Henderson: 1999.

REFERENCES

Augé M: 1995 *Non-places : introduction to an anthropology of supermodernity.* Translated by John Howe. London & New York, Verso.

Bertuglia, F.: 2002 'Le tipologie nella *città* diffusa' in Dal Pozzo *Fuori città , senza campagna*, pp. 100-133.

Clark, W.: 1998 *The California cauldron: immigrants and the fortunes of local communities*. New York, Guilford Press.

Cosgrove, D.: 1993 *The Palladian landscape: geographical change and its cultural representation in 16th century Italy.* State College PA, Penn State University Press.

Culver, L.: 2004 'The island, the oasis and the city: Santa Catalina, Palm Springs, Los Angeles and Southern California's shaping of American life and leisure' PhD Dissertation, University of California Los Angeles.

Dal Pozzo, L., ed: 2002 *Fuori città , senza campagna: paesaggio e progetto nella città diffusa*. Milano, FrancoAngelli.

Davis, M.: 1992 *City of quartz: excavating the future from Los Angeles*. London & New York, Verso.

DeLyser, D.: 2003 'Ramona memories: fiction, tourist practices, and placing the past in Southern California' *Annals, Association of American Geographers*, 93, pp. 888-908.

Detragiache, A. ed: 2003 *Dalla città diffusa alla città diramata*. Milano, FrancoAngeli.

Erbani, F.: 2002 'La *città* diffusa: cosi villette e capannoni diventano una megalopoli' *La Repubblica* 24/7/2002.

Fulton, W.: 2001 *The reluctant metropolis: the politics of urban growth in Los Angeles*. Baltimore and London, Johns Hopkins University Press.

Henderson, G - W: 1999 *California and the fictions of capital*. Oxford, Oxford University Press.

Katz, P. ed: 1994 *The new urbanism: toward an architecture of community.* New York, McGraw-Hill.

Ludoviana, F. ed: 1990 *La città diffusa*. Venezia, Daest.

McClung, W- A-: 2000 *Landscapes of desire: Anglo mythologies of Los Angeles*. Berkeley, University of California Press.

Mitchell, D.: 1996 *The lie of the land: migrant workers and the California landscape*. Minneapolis, Minnesota University Press.

Relph, E.: 1979 Place and placelessness London, Pion.
Soja, E.: 1996 *Thirdspace : journeys to Los Angeles and other real-and-imagined places.* Cambridge, Mass., Blackwell,
Terkenli, T. S: 2005 'New landscape spatialities: the changing scales of function and symbolism' *Landscape and urban planning,* 70, pp. 165-176. Thiene.it: 2004:'Veneto: riflessioni sul futuro' 3pp. http:www.thiene.it/forum   (12/3/2004)

ANDREW SLUYTER

TRAVELING/WRITING THE UNWORLD WITH ALEXANDER VON HUMBOLDT [1]

View of chinampa agriculture near town of Tlahuac south of Mexico City. Photo taken by Robert C. West, July 1947

The World Bank estimates that some twenty-five million Mexicans, a quarter of that country's population, live in extreme poverty and suffer chronic hunger. The causes of that horrible situation are theoretically controversial, even in general terms, but so long as development remains broadly defined as improvement in social and environmental well-being few would argue against the urgent need for policies that would develop Mexico and the other Latin American countries that Alexander von Humboldt traveled through at the beginning of the nineteenth century. The net result of six decades of international development effort, however, as measured from the founding of the World Bank in the 1940s, has been an increase rather than a decrease in the number of impoverished Mexicans and, more broadly, people throughout Latin America. Many, therefore, should agree that development theory requires

93

T.S. Terkenli & A-M. d'Hauteserre (eds.), Landscapes of a New Cultural Economy of Space, 93-116.

improvement so that it can begin to provide a basis for policies that actually work. And at least some should also agree—given globalization's increasingly undeniable entanglements of politics, economy, and culture—that such improvement should include the systematic integration of culture theory into a project thus far dominated by political-economy theory.

The concept of unworldment introduced by Terkenli in Chapter 1 might provide just the necessary analytic leverage to achieve such integration. Development, developed, undeveloped, underdeveloped, developing, and so on all have conceptual roots in political-economy, and derivative policy therefore emphasizes trade and governance issues. In contrast, unworldment and its putative variants—worldment, unworlded, underworlded, lesser worlded, worlding, and so on—derive from culture theory: not deterministic, categorical culture theory but processual culture theory (Sluyter, 1999). Unworldment signifies the processes through which places and their landscapes become unmade, through which sense of place dissipates along with awareness of involvement in the material/conceptual transformation of landscapes over many generations (Sluyter, 2001). Worldment would thus be the positive variant of unworldment, the cultural analog of economic development, and designate the multigenerational process by which people become native to a place (Jackson 1994).

This chapter addresses the potential of the concept of unworldment and its variants to provide a basis for the integration of culture theory into development theory and thereby help to explain why so many more Mexicans suffer extreme poverty now than half a century ago. Among culture theorists, Mary Louise Pratt (1992) assigns von Humboldt a major role in unworlding Latin America but emphasizes cultural processes to the near exclusion of political-economy. Her analysis thus remains largely incommensurable with development theory. To integrate culture theory and development theory, this chapter follows Pratt in her focus on von Humboldt but uses landscape, a phenomenon that incorporates the full spectrum of biophysical and social processes, to reveal the linkages between unworldment and undevelopment.

Specifically, what follows analyzes the Gulf lowlands of Mexico in relation to unworldment, undevelopment, and the interactions among von Humboldt, his texts, and the material/conceptual transformation of landscapes (Figure 1). The next section proceeds from an elaboration of von Humboldt's role in the unworldment of Latin America in general, to his travels from March 1803 to March 1804 through Mexico, then part of the colony called New Spain, to the major text to derive from that year of research: the *Political Essay on the Kingdom of New Spain*. Throughout, analysis emphasizes the linkages between unworldment and undevelopment. The subsequent section continues that emphasis but focuses analysis on material/conceptual transformations of Gulf lowland landscapes, moving from the precolonial to the colonial to the (post)colonial

VON HUMBOLDT: THEORY, TRAVEL, TEXT

Von Humboldt and Unworldment

According to Pratt (1992, pp. 111-143), the texts von Humboldt wrote on the basis of his 1799-1804 journey through Latin America, just before the Spanish empire was to begin fracturing into independent republics, contributed significantly to that region's unworldment.

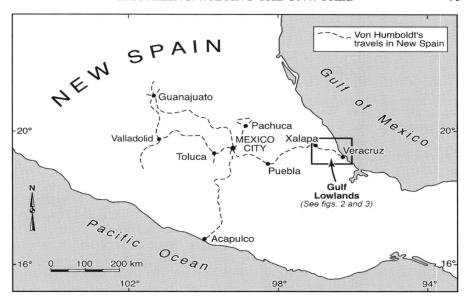

Figure 1. Von Humboldt's travels in New Spain and the location of the Gulf lowlands. Source: after Brand, 1959, fig. 1.

She concludes that his characterization of Latin America, particularly South America, as primordial nature gave scientific legitimacy to the pristine myth that had first emerged during the early colonial period, beginning with Columbus in 1492. That pristine myth maintains that the precolonial Americas were undeveloped, a hemispheric version of the more general myth of emptiness that came to pertain to many European colonies (Blaut, 1993; Denevan, 1992; Sluyter, 1999). In a process of unworldment, the sense of place of native peoples dissipated along with awareness of their involvement in the material/conceptual transformation of their landscapes over thousands of years. The places of Latin America thus became characterized as a New World that contrasted with the Old: one puerile, the other mature. At the end of the colonial period, von Humboldt so reinvigorated that pristine myth that it came to dominate (post)colonial Latin America.

> Even the label "New Continent" is revived, as if three centuries of European colonization had never happened or made difference. What held for Columbus held again for Humboldt: the state of primal nature is brought into being as a state in relation to the prospect of transformative intervention from Europe (Pratt, 1992, pp. 126-127).

Pratt does not much elaborate the linkages between von Humboldt's reinvigoration of the pristine myth and development, but the effect of such cultural unworldment on political-economy has indeed been profound (Sluyter, 1999). If the political-economic divide between the developed and the undeveloped emerged out of the colonial redistribution of global resources, labor, and capital, that same process involved the emergence of a cultural divide between the worlded and the unworlded (Adas, 1989; Said, 1979; Wolf, 1982). Colonial depopulation, truncation of oral histories, destruction of texts, and material/conceptual transformations of landscapes largely unmade the worlds of late precolonial Latin America (Sluyter, 2001). Europe became categorized as advanced, dynamic, and the source of everything good in contrast to backward, static, empty Latin America. That pristine myth has become so indurated as to resist erosion even by the accumulation of much contrary evidence (Turner and Butzer, 1992). Its development effects include the privileging of western institutions and technologies over native ones, the categorization of native peoples as being too traditional to participate actively in development, and the general conflation of economic development with cultural westernization (Sluyter, 2001).

Pratt's indictment of von Humboldt as instrumental to the unworldment of Latin America hinges, as do similar analyses more specific to Mexico (Florescano, 1994, pp. 203-204), on deconstruction of his writings in the context of his well-established dual status as one of the founders of modern science and as highly influential among the nineteenth-century elites of the Latin American republics. Regarding the nineteenth-century formation of the modern scientific disciplines, von Humboldt is seminal to geography and several other field sciences such as botany and anthropology (Livingstone, 1992, pp. 133-138). Some have even characterized science in the first half of the nineteenth century as "Humboldtian" because he provided a model for projecting the laboratory into the field through systematic measurement of widely distributed but related phenomena, thus revealing patterns and inducing general processes. Regarding his influence among the nineteenth-century elites of the (post)colonial Latin American republics, those of Mexico, for example, so admired him that upon independence the legislators of the State of Mexico declared him an honorary citizen, and upon his death in 1859 President Benito Juárez posthumously awarded him the Benemérito de la Patria (Mendoza Vargas & Azuela Bernal, 2003, p. 20; Miranda, 1962, pp. 106-107, 205-210; Pratt, 1992, pp. 111-113, 175-82; von Humboldt, 1980, p. 272).

Such textual analyses as Pratt's and Florescano's, however, necessarily remain partial because nobody, no matter their social or intellectual status, can unilaterally impose a concept about a place on a place. The (post)colonial reinvigoration of the colonial idea that the precolonial landscapes of the Americas were pristine must, by definition, relate to and in some way be consistent with other aspects, both conceptual and material, of those landscapes. In Mexico, local versions of the pristine myth emerged during the early colonial period through a process of material/conceptual landscape transformation (Sluyter, 1999). On logical grounds, similar processes must have occurred at the juncture between the colonial and the (post)colonial, and those processes must have involved the influential von Humboldt. His writings, as demonstrated by Pratt's and Florescano's analyses, were centrally involved in reinvigorating the pristine myth for Latin America in general and Mexico in specific. But those writings were based in good part on his observations of landscapes in South and Middle America, and therefore those writings partially derived from the colonial transformations of those landscapes and subsequently became involved in their (post)colonial transformations.

So the conclusions of literary scholars verge on idealism unless complemented with analysis of the long-term process of material/conceptual transformation of the landscapes von Humboldt traveled through and wrote about (Sluyter, 1997). The next two subsections therefore address, in turn, those travels and writings for Mexico in general.

Von Humboldt's Travels in Mexico

Together with Aimé Bonpland and Carlos Montúfar, von Humboldt landed at Acapulco on 22 March 1803, just seven years before the Grito de Dolores nominally began the War of Independence. He was thirty-three years old and had already spent nearly four years in the American tropics. In 1799, he had obtained permission from the Court in Madrid to conduct scientific research in its American colonies, an extraordinary accomplishment given that for three centuries the Spanish Crown had restricted entry mainly to its own subjects (Puig-Samper & Rebok, 2002; von Humboldt, 1980, pp. 248-250). Spanish officials had closely supervised previous exceptions, such as the French La Condamine expedition to Ecuador in the 1730s (Pratt, 1992, pp. 15-17). In contrast, von Humboldt had two passports in hand upon arriving in Acapulco, one from the Council of the Indies and the other from the First Secretary of State, both giving largely unrestricted and unsupervised freedom to travel and collect data (von Humboldt, 1815, pp. xi, 41-43). He nonetheless wrote Viceroy Iturrigaray from Acapulco to ask for permission to conduct scientific research in New Spain (Leitner, 2000; von Humboldt 1970, p. 142, 1980, pp. 107-108).

With the viceroy's encouragement, von Humboldt spent the next year traveling throughout central New Spain even though his instruments were in disrepair and he was eager to return to Europe (von Humboldt, 1970, p. 142). In fact, at least according to a letter he wrote on 29 April 1803 to the botanist Carl Ludwig Willdenow, von Humboldt remained in New Spain a full year rather than sailing from Veracruz during the summer of 1803 only because he feared the insalubrity of that port during the rainy season, when yellow fever gripped the Gulf lowlands (von Humboldt, 1980, p. 113). Nonetheless, he made an enormous number of firsthand observations and collected many unpublished statistics and maps in that year of research. By the beginning of 1804, he had accomplished enough to present a synthesis to Viceroy Iturrigaray, namely a manuscript entitled "Tablas geográficas políticas del Reino de Nueva España, que manifiestan su superficie, población, agricultura, fábricas, comercio, minas, rentas y fuerzas militares" (Archivo General de la Nación, Mexico City, Ramo Historia, vol.72, 2ª parte, exp. 24, ff. 271-294; von Humboldt, 1970, 1980, pp. 125-126). Also, by February 1804, the winter dry season had granted the Gulf lowlands their annual respite from yellow fever. And on 7 March 1804, von Humboldt, Bonpland, and Montúfar sailed from Veracruz for the United States by way of Cuba.

Published correspondence from the beginning and end of that year in New Spain demonstrates that while von Humboldt's purpose was in part basic scientific research, he also intended to contribute to development. His January 1804 letter to Viceroy Iturrigaray that accompanied the "Tablas geográficas políticas" illustrates how thoroughly von Humboldt believed that science should be applied as policy.

> In dealing with me regarding various works that I have carried out in this Kingdom, Your Excellency has deigned to hint that I should send him some interesting materials for the governance of these vast dominions. This hint for me has been an order with which I have complied so enthusiastically that my travels have had no other aim than to contribute with my findings to the public good that occupies no one in these parts more than Your Excellence (von Humboldt, 1980, pp. 125-126; author's translation of Spanish original).

That sentiment echoed the 1803 letter with which Viceroy Iturrigaray had responded to von Humboldt's request for a passport, which not only lauds his dedication to the sciences but also to "the welfare of humanity and other commendable aims" (von Humboldt, 1970, p. 142; author's translation of Spanish original). That desire to apply basic research to development, so clearly established while conducting the research in New Spain, became even more fully realized in the *Political Essay*.

Von Humboldt's Political Essay on Mexico

On returning to Europe, von Humboldt took up residence in Paris and began to prepare his journals for publication under the uniform title of *Voyage aux régions équinoxiales du nouveau continent fait en 1799, 1800, 1801, 1802, 1803 et 1804, par Alexandre de Humboldt et Aimé Bonpland*. Mainly due to financial constraints, the original publication plan changed several times, much confusing the project's history (Leitner, 2000; von Humboldt, 1815, pp. xvii-xxiv). Prepublication of fascicles, excerpts, parallel editions in pricey and economical formats, authorized and unauthorized translations, and non-uniform titles and inconsistent title pages all add further confusion to the Humboldtian bibliography. Nonetheless, despite having moved from Paris to Berlin in 1829, he had published thirty-four volumes of *Voyage aux régions équinoxiales* by 1838. His attention then turned to *Kosmos*, the first volume of which appeared in 1845.

Several volumes of *Voyage aux régions équinoxiales* treat Mexico to some degree. *Recueil d'observations astronomiques, d'opérations trigonométriques et des mesures barométriques*, however, merely records the instrumental observations (von Humboldt, 1810a). *Vues des Cordillères et monumens des peuples indigènes de l'Amérique* juxtaposes images and descriptions of precolonial art and architecture with those of landscapes, some of them Mexican (von Humboldt, 1810b). But while the form of *Vues des Cordillères* suggests much about von Humboldt's ontology, such as his belief that the precolonial peoples of the Americas were part of nature, it contains little of direct relevance to the Gulf lowlands.

Only the *Political Essay on the Kingdom of New Spain*, the volumes of which form the third part of *Voyage aux régions équinoxiales*, directly focuses on what was soon to become the republic of Mexico. Humboldt originally published the *Political Essay* as a series of seven fascicles, with the first one appearing in 1808 and the entire set collected in 1811 into two volumes and an atlas, a year after the Grito de Dolores and a decade before the final victory over Spanish sovereign power that ushered in (post)colonial Mexico. The uniform title was *Essai politique sur le royaume de la Nouvelle-Espagne avec un atlas physique et géographique, fondé sur des observations astronomiques, des mesures trigonométriques et nivellemens barométriques* (von Humboldt, 1811). It incorporated the "Tablas geográficas políticas" he had presented to Viceroy Iturrigaray in 1804 but greatly expanded the non-tabular sections with material from his journals.

Perhaps for that reason he never published the planned fourth volume of the *Relation historique du voyage aux régions équinoxiales. . .*, which comprises the first section of the first part of *Voyage aux régions équinoxiales* and is usually translated into English as the *Personal Narrative of Travels to the Equinoctial Regions of the New Continent, During the Years 1799-1804* (e.g., von Humboldt, 1815, 1941). The three published volumes of the *Relation historique* narrate the journey in South America and Cuba. The planned fourth volume, treating the journey in Mexico, was apparently abandoned while in press (von Humboldt, 1941, pp. vii-viii). Pratt (1992, p. 129) believes that von Humboldt found personal travel narratives so "repugnant" that he idealistically cancelled the printing at the last minute. But maybe the *Essai politique* had simply grown so popular that von Humboldt and the publisher pragmatically decided that a fourth volume of the *Relation historique du voyage* on Mexico would be redundant and lose money. And the *Essai politique* certainly was popular, resulting in a revised French edition of 1825-27 and numerous translations, including an English edition published by Longman in 1811, a Spanish edition of 1822, and an Italian edition of 1827-29 (Leitner, 2000).[2]

The *Political Essay* was organized into a Geographical Introduction and six books, themselves divided into a total of fourteen chapters. The Geographical Introduction details the sources for the many maps and other figures, ranging from tables of von Humboldt's own astronomical and barometric observations to lists of manuscript maps and accounts of landscape sketches. Book 1 treats the land of New Spain. Its three chapters describe the colony's territorial extent and political divisions, its coastlines, and physiographic aspects such as soils and climate. Of note, it describes transcontinental routes, the potential for an inter-oceanic canal, and the now classic relationship between climate and elevation: tierras calientes, templadas, frías, and heladas. The four chapters of Book 2 concern the people of New Spain: population change; epidemics, famines, dangerous working conditions, and other impacts on population growth; native ethnic groups; and other ethic groups (European, African, and mixed). It particularly emphasizes the growth in population in the decade since the Revillagigedo census of 1793, the state of the native population, and social inequity in relation to ethnicity. Book 3 consists of a single chapter that describes each of the fifteen territories within New Spain, focusing on surface area, population, and population density. The intendancy of Mexico receives as many pages as all of the other fourteen territories combined. The fourth book treats the prospects for agricultural and mining development in three chapters. Chapter 1 deals with the different crop plants and their many varieties, Chapter 2 with the different animals; and Chapter 3 with the different minerals and the methods of mining them. It describes at length the details of different mining techniques, reflecting von Humboldt's experience as a mine manager in Silesia. Book 5 has a single chapter that deals with the prospects for manufacturing and commerce. Of note, it emphasizes that a lack of factories necessitates the importation of goods. That situation, concludes von Humboldt, reduces the amount of capital available to develop a domestic manufacturing sector. The two chapters of Book 6 describe, in turn, state revenues and military defense. Lengthy notes, supplements, and an index follow the sixth book. In all, even without including the detailed index, the four-volume Longman edition of 1811, which includes small reproductions of some of the plates from the atlas of the 1811 French edition, runs to 1,833 pages—quite an *essay*.

Besides being a turgidly wonderful compilation of detailed data on late colonial New Spain, the *Political Essay* is, according to Donald Brand (1959, p. 123), "the first modern regional economic geography. . . , concerned primarily with the sources of wealth and their distribution and utilization." As such, the *Political Essay* is seminal to the emergence of modern economic development planning in Mexico and elsewhere, a canonical text among country-level development reports. For example, the opening paragraph of the World Bank's 2001 development report for Mexico states that, "It has been the privilege of the World Bank to provide incoming Presidential Administrations in its client countries with a comprehensive account of its diagnoses and policy recommendations for the sectors that contribute to the client's development path" (Oliver, Vinh, & Giugale, 2001, p. 1). The *Political Essay* begins similarly, von Humboldt (1966, vol. 1, p. 1) stating his purpose as "an investigation of the causes which have had the greatest influence on the progress of the population and national industry." In a letter to Charles IV of Spain, von Humboldt (1966, vol. 1, p. xvii) is similarly explicit regarding his purpose, ending with the following sentence, "How can we displease a good king, when we speak to him of the national interest, of the improvement of social institutions, and the eternal principles on which the prosperity of nations is founded?"

Moreover, the form of the *Political Essay* parallels the political-economy conceptual framework that became so characteristic of orthodox development theory. The analytic starting point is the so-called natural and human resource base and its geographical variation, which is exactly how von Humboldt begins the *Political Essay* with Books 1 through 3. Discussion of the recommendations for development then proceeds sector by economic sector, just as von Humboldt addresses the primary through tertiary sectors in Books 4 and 5. That conceptual framework then culminates with an analysis of the social capacity to implement those recommendations, exactly in the way that von Humboldt concludes the *Political Essay* with Book 6 by providing an analysis of the capacity to govern New Spain. Emphasis might have shifted to developing the quaternary sector and building the capacity of civil society, but the structure of the *Political Essay* nonetheless parallels that of (post)colonial development reports.

Yet, as anyone but an idealist would agree, von Humboldt based his *Political Essay* in good part on observations of landscapes, and therefore that text partially derives from the colonial material/conceptual transformations of those landscapes and subsequently became involved in their (post)colonial transformations (Sluyter, 1997). Understanding the linkages among von Humboldt, unworldment, and undevelopment therefore requires analysis of specific landscapes in relation to the text of the *Political Essay* (Sluyter, 2001). The next section therefore focuses analysis on the Gulf lowland landscapes, proceeding from the precolonial to the colonial to the (post)colonial.

TRAVELING/WRITING GULF LOWLAND LANDSCAPES

Von Humboldt traveled across the Gulf lowlands in early 1804 on his way from Xalapa to the port of Veracruz, where he, Bonpland, and Montúfar embarked for the USA by way of Cuba. By then, that region's landscapes had undergone nearly three centuries of unworldment and undevelopment (Figure 2).

Developed Precolonial World

In contrast, when Hernán Cortés arrived at Veracruz in 1519, he and the other conquistadors had encountered densely settled and productively developed landscapes that manifested a precolonial world (Sluyter, 2002, pp. 35-60). Zempoala,

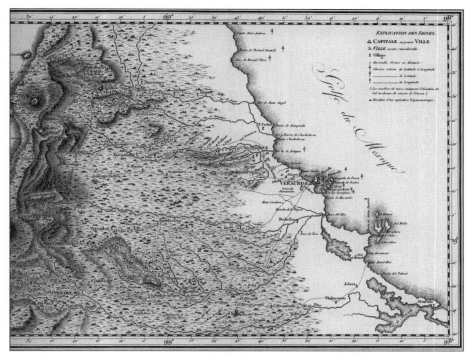

Figure 2. Von Humboldt's map of the Gulf lowlands, excerpted from his map of New Spain.
Source: von Humboldt, 1811, vol. 3, plate 9.

a city of a hundred thousand inhabitants, occupied the top of a settlement hierarchy with a total population of some half a million in an area of about 5,000 square-kilometers (Figure 3). Those people had developed several types of intensive agriculture attuned to the highly seasonal precipitation regime of the subhumid climate, including sloping-field terracing, intensive wetland agriculture and, probably, some canal irrigation systems of limited extent.

The terraces occupied the piedmont slopes, occurring in complexes of hundreds of hectares. The chronology of those fields and their precise extent both remain elusive, but farmers seem to have constructed them by clearing fieldstones into lines parallel to slope contour. Beyond field clearance, the terraces functioned to manage soil moisture, essential given the climate. The principal food crop was maize, with cotton and agave grown for fiber.

Intensive wetland agriculture covered several thousand hectares of the belt of backswamps just inland from the cordon of dunes along the coast. Although the precise extent and chronology of the wetland fields also both remain uncertain, farmers seem to have constructed the fields by ditching into and mounding above ground surfaces that seasonally intersected the water table. The most general function involved regulation of soil moisture in the rooting zone to allow maize cropping as early in the dry season as possible yet retain enough water for splash and subirrigation well into the dry season.

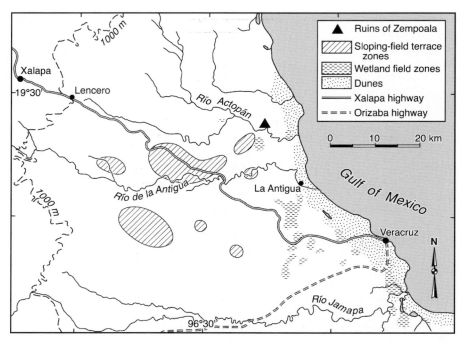

Figure 3. The Gulf lowlands with locations of places, routes, and field vestiges mentioned.
Source: after Siemens, 1990, fig. 1; Sluyter, 2002, fig. 2.6.

Colonial Unworldment and Undevelopment

With rapid and near total depopulation during the early colonial period, on the scale of ninety-nine percent by the end of the sixteenth century, the Gulf lowlands underwent a process of material/conceptual landscape transformation that resulted in drastic undevelopment and unworldment (Sluyter, 2002, pp. 143-185). The interaction of precolonial land use and climate had created a matrix of cultural savanna with patches of settlement and agriculture by the early sixteenth century. That material landscape patterning together with landscape concepts and laws diffused from Spain through the Antilles constituted the parameters for the Spanish vision of a pastoral landscape based on an economy of livestock ranching. Over the course of the sixteenth century, native depopulation due to epidemics of introduced diseases created a moribund landscape, thickets invading former fields and settlements before the livestock. As the ranchers accumulated space and increasingly occupied the landscape after mid century, they preempted the recovery of the surviving native population.

Epidemiological and old-field succession processes thus resulted in a material landscape pattern that together with the conceptual landscape pattern inherent in such categories as "wasteland" mediated further landscape transformation. The recategorization as wasteland of both moribund and fallow fields belonging to native communities obscured the native labor that had developed those landscape patches, that would have provided the basis in Spanish law to prevent dispossession, and that would thus have made possible the recovery of native population and reversal of old-field succession. In effect, a positive feedback loop linked the material and conceptual transformations to render them inexorably unidirectional, resulting in the taken-for-granted belief that an undeveloped, unworlded, pristine landscape was progressing toward a productive, developed landscape when, in fact, the opposite was occurring.

By the late colonial period, when von Humboldt traveled the royal highway from Xalapa to Veracruz, the piedmont and wetlands had become well established as poles in a system of regional transhumance. Tens of thousands of cattle grazed the wetlands during the dry season, retreating to the higher and drier piedmont with the onset of the rains. Settlement was sparse, with one nineteenth-century observer noting that neither "towns nor villages are found in these extensive districts, but merely here and there the solitary farms of the cattle-proprietors, or of the herdsmen" (Sartorius, 1961, pp. 9-10). The precolonial world had disappeared along with its native inhabitants and the knowledge of how they had over many generations productively developed the landscapes of their place. A local version of the pristine myth prevailed.

Von Humboldt, far from refuting that pristine myth at the end of the colonial period, reinvigorated it by not acknowledging the dense precolonial settlement and productive development of the Gulf lowlands. He did acknowledge that precolonial population was higher and land use more intensive than in 1804.

> Chalchiuhcuecan (Vera Cruz), all the country from the river of Papaloapan (Alvarado to Huaxtecapan) [Papaloapan (Alvarado) jusqu'á Huaxtecapan,], was better inhabited and better cultivated than it now is (von Humboldt, 1966, vol. 2, p. 254).

But relative to the highlands, he characterized the lowland population as low in both precolonial and colonial times.

> However, the conquerors found as they ascended the table land [plateau] the villages closer together, the fields divided into smaller portions, and the people more polished [policé] (von Humboldt, 1966, vol. 2, p. 254).

> Collected together on a small extent of territory, in the center of the kingdom, on the very ridge [plateau] of the Cordillera, they have allowed the regions of the greatest fertility and the [sic] nearest to the coast to remain waste and uninhabited. . . . In this the Spanish conquerors have merely trod in the steps of the conquered nations. The Aztecs, originally from a country to the north of the Rio Gila, perhaps even emigrants from the most northern parts of Asia, in their progress towards the south never quitted the ridge of the Cordillera, preferring these cold regions to the excessive heat of the coast (von Humboldt, 1966, vol. 1, pp. 89-90).

If indictment requires opportunity, von Humboldt certainly did not lack evidence to refute the pristine myth, both textual and landscape evidence (Sluyter, 2004). In terms of texts that demonstrated the dense settlement and productive development of the precolonial Gulf lowlands, he spent much of 1803 "in the intellectual and scientific communities of Mexico City, where he studied existing corpuses on natural history, linguistics, and archaeology" (Pratt, 1992, p. 136). The primary texts of that body of literature included eyewitness accounts of late precolonial landscapes written by conquistadors. They also included compilations of precolonial and early colonial histories that members of the clergy collected and codified during the sixteenth century, when a greater proportion of the precolonial corpus had still escaped the destruction and depopulation of the colonization process.

The most pertinent of the eyewitness accounts by conquistadors are the *Historia Verdadera de la Conquista de la Nueva España* by Bernal Díaz del Castillo (1986), the *Cartas de Relación* by Hernán Cortés (1988), and *La Conquista de México* by Francisco López de Gómara (1987), Cortés's biographer. All contain evidence that refutes the pristine myth for the Gulf lowlands, yet von Humboldt assiduously avoided that evidence even while using those same three texts to support other points he makes in the *Political Essay*. Díaz's reaction upon riding into Zempoala for the first time provides one example of the type of evidence von Humboldt ignored.

> [Entering] among the houses, on seeing such a large city, and having seen no other larger, we greatly admired it, and how it was so luxuriant and like a garden, and so populous with men and women, the streets full of those who had come out to see us (Díaz del Castillo, 1986, p. 76).

Similarly, von Humboldt read some of the sixteenth-century compilations of precolonial and early colonial histories, which preserved at least some of the precolonial corpus of codices and oral histories. The Codex Mendoza, a copy of a register of tribute levied on the provinces of the Aztec empire, provides one example that he drew on, even reproducing some of its pictographs in *Vues des Cordillères* (von Humboldt, 1810b, pp. 284-291 and plates 58-59, 1966, vol. 2, p. 18). The Codex Mendoza includes the tribute for the province of Cuetlaxtlan, which encompassed the port of Veracruz and environs, the Spanish glosses noting an annual tribute of 6,720 loads of cotton mantles, 200 loads of cacao, and various luxury items. Such evidence should have made clear to von Humboldt, at least in qualitative terms, the intensive cultivation and dense settlement of the late precolonial Gulf lowlands.

Monarquía Indiana by Juan de Torquemada (1969), written beginning in 1592 and first published in 1615, provides another example of such textual evidence (Franch, 1973). Von Humboldt (1966, vol. 2, pp. 18, 74) cited both *Monarquía Indiana* and Francisco Clavigero's 1780-81 *Storia Antica del Messico*, which itself reiterates much of Torquemada (Clavigero, 1780-81, 1917; Ronan, 1973). Yet von Humboldt ignored the evidence that both texts provide regarding the precolonial Gulf lowlands being anything but pristine, such as an account of the origins of Zempoala and a description of its environs as densely populated (Clavigero, 1917, vol. 2, p. 25; Torquemada, 1969, vol. 1, p. 278). Torquemada (1969, vol. 1, pp. 251, 396; author's translation of Spanish original), in fact, claimed that Zempoala was a grand city with a population on the order of "twenty-five to thirty thousand *vecinos*," large buildings, broad avenues, and many houses with lush gardens—"altogether appearing a delightful paradise." Since *vecinos* refers to heads of households, the total population would have been 4.5 times as great, some 112,500 to 135,000 (Sluyter, 2002, p. 44). Other such references suggest a late precolonial population of some half a million for the region as a whole (Sluyter, 2002, p. 47; Torquemada, 1969, vol. 1, p. 522).

Clearly, using such texts to reconstruct precolonial populations requires awareness and correction for their inherent biases (Sluyter 2002, pp. 41-47). Among the primary texts, Cortés might have estimated population rather liberally to exaggerate his accomplishments and the potential for colonization. Torquemada inflated even Cortés's estimates. Díaz, in contrast, explicitly attempted to counter such exaggeration with his *Historia Verdadera*.

Moreover, among von Humboldt's secondary sources, some clearly bastardized the primary texts while others were explicitly engaged in polemics related to early Creole patriotism. Torquemada, for one, praised precolonial antiquity and Mexican nature, a patriotism that would became the reasoned, Enlightenment nationalism of the later eighteenth century (Florescano, 1994, p. 187). Clavigero, for example, on whom von Humboldt relied more than any source other than Cortés, drew disproportionately on Torquemada in order to counter the eighteenth-century thesis that the Americas were naturally inferior to Europe. That so-called *querelle d'Amérique* stemmed from Buffon's claims about the poverty of American nature and society relative to Europe but became popularized through such publications as Cornelius de Pauw's 1768 *Recherches Philosophiques sur les Américains* and William Robertson's 1777 *History of America*, both of which Clavigero directly attacked (Clavigero, 1917, vol. 2, pp. 431-449; Florescano, 1994, pp. 189-191; Glacken, 1967, pp. 680-685; Pratt, 1992, p. 120).

Yet if clear cases of polemical hyperbole caused von Humboldt to doubt the veracity of the available textual evidence, he should not have so entirely ignored those sources when they contradicted the pristine myth for the Gulf lowlands at the same time that he relied so much on them to support some of his other conclusions in the *Political Essay* (Sluyter 2004). Moreover, the references to the dense settlement of the precolonial Gulf lowlands should certainly have encouraged him to look for the presence or absence of corroborating landscape evidence along the highway between Xalapa and Veracruz.

Yet he did not visit even the ruins of Zempoala, a precolonial city of some 100,000 people. Francisco Lorenzana's 1770 *Historia de Nueva España*, which von Humboldt otherwise cites repeatedly in the *Political Essay*, roughly provides the location of Zempoala's ruins (Lorenzana, 1980, p. 49; von Humboldt, 1966, vol. 1, p. xxxi). An excursion of less than twenty kilometers off the highway would have allowed von Humboldt to visit the ruins of a large precolonial city that, because of rapid depopulation and a relative lack of fighting during the early sixteenth century, remained well preserved compared to the city of Tenochtitlán, which the Spaniards had demolished and rebuilt as Mexico City. Yet von Humboldt did not even mark Zempoala's location on his map of the Gulf lowlands and thereby helped to obscure its location until rediscovery circa 1880 by Estefania Salas (Strebel, 1883).

Moreover, von Humboldt acknowledged nothing of the vestiges of precolonial intensive agriculture along the highway. Certainly, at ground level, the vestigial wetland fields appear as no more than subtle topographic and vegetational variations. So despite the highway running right beside several complexes of wetland fields, they disappeared from the conceptual landscape between 1560—when Lucas Hernández noted "a small lake which appears in the rainy season. . . and marshes ditched straight southward" (Archivo General de la Nación, Mexico City, Ramo Mercedes, vol. 15, ff. 191v)—and the 1970s, when Alfred H. Siemens (1980) noted their characteristic vegetational patterning while flying into the Veracruz airport. Moreover, von Humboldt (1966, vol. 4, p. 154) seems to have left the royal highway where it crosses the Río de la Antigua and instead proceeded to Veracruz by way of the town of La Antigua and the coastal route. He might therefore have never even been within sight of a complex of vestigial wetland fields, difficult in any case to distinguish from ground level.

While an aerial perspective also makes the sloping-field terraces more evident, other visitors to the Gulf lowlands, also German and also limited to a ground-level view of the landscape, reported the vestiges just a few years after von Humboldt failed to do so. Carl Sartorius arrived in Mexico in 1824 and published *Mexico: Landscapes and Popular Sketches* in 1858, providing a clear description of the precolonial terraces.

> When the tall grass is burnt down, we can see that the whole country was formed into terraces with the assistance of masonry, everywhere provision had been made against the ravages of the tropical rains; they were carried out on every slope, descending even to the steepest spots, where they are often only a few feet in width (Sartorius, 1961, p. 10).

Brantz Mayer (1847, p. 11) and Hugo Fink (1871, p. 373), two other nineteenth-century travelers from Germany, similarly reported the vestigial terracing.

So von Humboldt, the quintessential field scientist, should certainly have recognized those terrace vestiges, as well as the extensive ruins of precolonial settlement that Sartorius also reported: "On the dry flat ridges the remains of large cities are found, forming for miles regular roads" (Sartorius, 1961, p. 10). After all, von Humboldt crossed the lowlands in late February, when the dry season was far enough advanced for the deciduous vegetation to have lost its leaves and the grass to have withered. The highway passed through extensive complexes of vestigial terraces. And the ranchers would have been burning the range, their fires exposing the characteristic rows of stones running along slope contours.

Instead, von Humboldt not only accepted the pristine myth for the Gulf lowlands, he was generally negative about their agricultural potential. He characterized the precolonial Gulf lowlands as populated, albeit sparsely relative to the highlands, and undeveloped relative to their potential. He characterized the colonial Gulf lowlands as even more sparsely populated than precolonially and undeveloped relative to their potential. And he implied that since native peoples and Spanish colonizers had failed to develop the Gulf lowlands beyond extensive cattle ranching, in his view a waste of their potential, European colonists not subject to what he believed to be lassitude caused by the tropical climate would have to implement intensive agriculture (von Humboldt 1966, vol. 2, pp. 253-256, vol. 3, p. 101, vol. 4, pp. 154-156).

Less clear than the lack of rigor in the use of textual sources and field observations that resulted in such conclusions, are the proximate causes of that lack of rigor. Pratt (1992, pp. 136-137) contends that because von Humboldt spent so much time immersed in the scholarly community of Mexico City, his writings heteroglossically incorporated the voices of his Creole informants, autoethnographically reflecting their representations of themselves. He thereby transculturated those representations into existing European representations of New Spain (also see Florescano, 1994, pp. 204-205). And following independence, (post)colonial Mexican elites reimported their own ideas as scientific, "*European knowledge* whose authority would legitimate Euroamerican rule" (Pratt, 1992, pp. 136-137; italics in original). The republican elites then used ideas such as the pristine myth to continue to disempower those whom they continued to colonize, namely native peoples.

The actual process involved in such autoethnography related to, perhaps most basically, the overwhelming emphasis on the Basin of Mexico (the location of Mexico City) and the Aztecs in the texts von Humboldt relied on, such as the *Cartas de Relación*, Codex Mendoza, *Monarquía Indiana*, and *Storia Antica del Messico*. That bias obscures the more scattered and limited information on the Gulf lowlands. Perhaps von Humboldt did, as Pratt suggests, reach his conclusions on the basis of what Mexican scholars told him and merely selectively skimmed the primary texts for corroborating rather than contradictory evidence. Perhaps he even entrusted Montúfar, who was fluent in Spanish, to locate such passages (Pratt, 1992, p. 240). Ferreting out the brief passages relevant to the Gulf lowlands requires long hours of careful reading, and systematic content analysis of the texts indicates that von Humboldt never took the time to be that careful, either in Mexico City or later in Paris (Sluyter 2004).

Yet von Humboldt's reputation as an avid field scientist and keen observer suggests that he should have used landscape evidence as an independent source to support or refute the ideas of his Creole informants. Careful field observation would have mitigated the autoethnographical tendencies of research based on texts and informants. As with the textual evidence, the process involved again seems to relate to not taking the time for careful, rigorous observations. Von Humboldt and his companions took only two days to travel from Xalapa to Veracruz, seemingly too preoccupied with the threat of yellow fever to carefully observe lowland landscapes before hurriedly departing New Spain (Sluyter, 2004).

(Post)colonial Misdevelopment and Misworldment

People began to resettle the Gulf lowlands in number only after the agrarian revolution that ended the dictatorship of Porfirio Díaz, when von Humboldt's twentieth-century scions, the (post)colonial development planners, would repeat his error regarding the precolonial landscape (Sluyter, 2002, pp. 191-202). In 1910, as the Porfiriato (1876-1911) neared its end, haciendas focused on ranching controlled most of the Gulf lowlands, with 78% of the population of 62,589 living in and around the port of Veracruz itself. By 1990, land redistribution had broken up some of the large estates and population had increased to 776,066, with some 61% in and around the port of Veracruz and around 250,000 spread over the rest of the region and beginning to approach precolonial population densities.

That process of resettlement has hinged on a radical hydrological transformation that epitomizes the egregious failures of such projects in Mexico, the birthplace of the Green Revolution as well as (post)colonial development more generally (Sluyter, 2002, pp. 196-199). Over the course of the late 1930s through the early 1970s, development planners created a network of dams, pumps, irrigation canals, and drainage ditches to convert the colonial pastoral landscape into an agricultural one, at least in part. That agriculture, however, is dominated by exotic biota, domesticated in Asia and brought to the Americas by Europeans for export as commodities. Native food crops such as maize and beans to feed Mexicans occupy mere landscape interstices. The planners thought of themselves as engaged in a heroic battle against the limitations of climate, topography, hydrology, and so-called traditional culture

(Siemens, 1998, p. 213; Skerritt Gardner, 1993, p. 17). In their discourse, development would require that ranching give way to intensive, irrigated agriculture just as the pristine precolonial landscape supposedly had given way to ranching during the colonial period. The "nasty tropical swamps" were "too wet" and "poorly drained," the encompassing coastal plain and piedmont "too dry" for much of the year. The planners would homogenize the seasonality of climate and hydrology through irrigation and drainage.

On the one hand, in ignoring the potential of precolonial native ecologies such as sloping-field terracing and intensive wetland agriculture, the planners accepted von Humboldt's characterization of the precolonial Gulf lowlands as relatively pristine. On the other hand, they also ignored those like Sartorius who had recognized precolonial development.

Pratt (1992, pp. 133-135) provides a critical insight into how a process she refers to as "archaeologization," and which she credits in large part to von Humboldt, rationalized the contradiction between acknowledging precolonial development and ignoring it as a model for (post)colonial development. She notes that in von Humboldt's writings on the precolonial peoples of South America,

> the links between societies being archaeologized and their contemporary descendants remain absolutely obscure, indeed irrecoverable. . . . The European imagination produces archaeological subjects by splitting contemporary non-European peoples off from their precolonial, and even their colonial, pasts. To revive indigenous history and culture as archaeology is to revive them *as dead*. The gesture simultaneously rescues them from European forgetfulness and reassigns them to a departed age (Pratt, 1992, p. 134).

Archaeologization thus glorifies precolonial native peoples to the exclusion of living ones (Florescano, 1994, pp. 192-195; Pratt, 1992, pp. 136-137).

Although Pratt credits von Humboldt with legitimizing archaeologization as scientific, she recognizes that he derives the method from the nationalistic Creole scholars of the late colonial period, such as Clavigero. In the same way that von Humboldt autoethnographically incorporated so much else into the *Political Essay*, giving the biases of Creole nationalists the credibility of science, he systematized, improved on, and legitimized their method of archaeologization.

The first step in that method disparages contemporary natives.

> The American Indians, like the inhabitants of Hindostan, are contented with the smallest quantity of aliment on which life can be supported, and increase in number without a proportional increase in the means of subsistence. Naturally indolent, from their fine climate and generally fertile soil, they cultivate as much maize, potatoes, or wheat as is necessary for their own subsistence, or at most for the additional consumption of the adjacent towns and mines (von Humboldt, 1966, vol. 1, p. 119).

Because of the lack of evidence for such "natural indolence," he did not espouse such negative conclusions categorically, however, typically employing the trope of self-critique and restraint to ensure his authority as an objective scientist.

> I deliver this opinion, however, with great reserve. We ought to be infinitely circumspect in pronouncing on the moral or intellectual dispositions of nations from which we are separated by the multiplied obstacles which result from a difference in language and a difference of manners and custom (von Humboldt, 1966, vol. 1, pp. 170-171).

Having thus established the, at least, suspicion that living native peoples are inherently unproductive, the second step in the logic of archaeologization attributes the accomplishments of precolonial native peoples to diffusions from the so-called Old World.

> I do not mean to discuss here what the Mexicans were before the Spanish conquest; this interesting subject has been already entered upon in the commencement of this chapter. When we consider that they had an almost exact knowledge of the duration of the year, that they intercalated at the end of their great cycle of 104 years with more accuracy than the Greeks, Romans, and Egyptians, we are tempted to believe that this progress is not the effect of the intellectual development of the Americans themselves, but that they were indebted for it to their communication with some very cultivated nations of central Asia (von Humboldt, 1966, vol. 1, pp. 158-159).

Given the lack of evidence for such pre-Columbian diffusions, though, scientific archaeologizers take a third logical step to ensure the dominance of western ideas over native ones. That final step characterizes living natives as so degraded that they bear no resemblance to their accomplished precolonial ancestors. In that view, even if precolonial Mexicans had any knowledge that could contribute to (post)colonial development, living natives have not inherited that knowledge.

> To give an accurate idea of the indigenous inhabitants of New Spain, it is not enough to paint them in their actual state of degradation and misery; we must go back to a remote period, when, governed by its own laws, a nation could display its proper energy; and we must consult the hieroglyphical paintings, buildings of hewn stone, and works of sculpture still in preservation, which, though they attest the infancy of the arts, bear, however, a striking analogy to several monuments of the most civilized people (von Humboldt, 1966, vol. 1, p. 140).

Such archaeologization became a prime method for (post)colonial elites to become non-European while at the same time accelerating the westernization of their newly independent republics (Pratt, 1992, p. 175). Their nationalistic indigenist rhetoric, or *indigenismo* reconciled retention of the pristine myth with glorification of the precolonial past by celebrating Aztec military heroes and the monumental architecture and art of the long-gone Classic Maya—albeit always measured against western benchmarks such as those of Classical Antiquity (Sluyter, 2002, p. 201). Thus the Maya become the Greeks of the New World and the Aztecs become the Romans. Meanwhile living natives and their land-use practices became categorically "traditional," supposedly trapped in a "culture of poverty"—with "race of poverty" lurking in the conceptual shadows—that makes them too conservative to actively participate in economic development models that focus on the diffusion and adoption of western institutions and technologies (Lewis 1966). The vestiges of precolonial agriculture, just beyond the pale of touristic ruins, remain largely unseen, unsung, and irrelevant.

Sartorius—scientist, entrepreneur, and promoter of German settlement in newly independent Mexico (Mentz, 1990, pp. 22-45)—provides an example. He believed that the vestiges of precolonial terracing demonstrated that the soil and climate of the Gulf lowlands would support cultivation of tropical commodities such as coffee and sugar cane, that the ranching established during the colonial period wasted that potential, and that immigration by westerners was necessary to realize that potential. In order to reconcile the contradiction inherent in using evidence of precolonial

development to promote (post)colonial development by westerners rather than
natives, he ascribed the terrace vestiges to an antiquity so ancient, and perhaps even
to Biblical migrants, as to lose all relation to living natives

> Nevertheless this region has a peculiar charm for men of an enquiring turn. Traces of
> extinct tribes are here met with, of a dense agricultural population, who had been
> extirpated before the Spaniards invaded the country. . . . All is now concealed by trees or
> tall grass; for many miles scarcely a hut is built, where formerly every foot of land was
> as diligently cultivated as the banks of the Nile or the Euphrates in Solomon's time. We
> know not whether a plague or hunger, or warlike tribes from the North, or some great
> convulsion of nature destroyed the numerous population, indeed we have not the
> slightest clue, which would enable us to decide to what people these relics of great
> industrial activity belong (Sartorius, 1961, p. 10).

Apparently late precolonial peoples and their living descendants were incapable
of having developed such a productive landscape, and pre-Columbian diffusions
from the West might well explain the vestiges of dense settlement and intensive
agriculture.

> An impartial consideration and observation of the Indians during many years forced me
> to the conclusion: that, according to their bodily organization, they are incapable of so
> high a degree of intellectual development as the Caucasian race. . . . The religious
> systems of the Incas and Aztecs, their knowledge of astronomy, works of art, and
> mechanical labours for the purposes of every-day life, are the result of their powers of
> understanding, of the undeniable imitative talents of the whole race. . . . As yet we know
> not whether influences from the east may not have sown the first seeds of civilization
> (Sartorius, 1961, p. 64).

(Post)colonial Reworldment and Alternative Development

Given that centuries-long process of unworldment, the (post)colonial
development planners who followed von Humboldt and Sartorius easily ignored the
potential of precolonial ecologies to inform development and, instead, implemented
extreme westernization, a policy that on the basis of its deleterious results had
already by the 1970s begun to seem like misdevelopment. By then it had become
clear that development theorists' faith in westernization had not improved social
well-being and, moreover, had made society subject to environmental degradation on
an unprecedented, global scale (Meadows et al., 1972). That double failure
stimulated alternative development models under the general rubric of sustainability.
One of them focuses on heterogeneous native ecologies—the ecological knowledges
and practices of the very cultures that the West colonized and long dismissed as too
traditional to participate in development—as the major source of alternatives to
homogenizing westernization (Sluyter & Siemens, 2004a).

In the Gulf lowlands, therefore, the contemporaneous recognition of the long-
elided vestiges of precolonial intensive wetland agriculture stimulated an attempt by
INIREB (Instituto Nacional de Investigaciones sobre Recursos Bióticos) to re-
implement that native ecology. In 1968, Siemens observed wetland patterning from a
small plane over Campeche, 500 kilometers east of Veracruz, and recognized the
significance (Sluyter, 1994). Over the 1970s, he and others made many more such
sightings to the east and west, revealing thousands of hectares of wetland fields along
the flanks of the Yucatan peninsula and in the environs of Veracruz. To

archaeologists, those vestiges suggested that lowland precolonial population densities had been much higher than previously believed. To INIREB, now defunct but then headquartered in Xalapa, intensive wetland agriculture became a primary component in a plan for the sustainable development of Mexico's tropical lowlands (Gómez-Pompa et al., 1982).

In INIREB's view, that native ecology promised to alleviate the situation of many lowland farmers who had access only to wetlands that were unsuitable for Green Revolution crops, which tend to require level, well-drained land suitable for irrigation. If the many wetlands of the Gulf lowlands could become productive components of integrated farms that produced high yields—high per unit of area, not necessarily high per unit of labor—of food crops, then rural people would be able to significantly increase their well-being and not be forced to migrate to a desperate existence in the burgeoning cities. Such integrated farms would, in INIREB's vision, emerge from the wetlands that misdevelopment had marginalized. And they would do so not by draining them but by modifying them in a way that sustained their hydrology and biodiversity.

Yet while growing recognition of the vestiges of precolonial wetland agriculture in the lowlands stimulated INIREB's project, the *chinampas* of the highland Basin of Mexico provided the model for construction and function. The ostensible reason for the application of a highland native ecology to the lowlands was that scientists have struggled to understand the construction and function of the lowland wetland fields because none remain extant (Sluyter, 1994). Clearly they were a way of manipulating soil moisture in places otherwise too wet to crop, at least for part of the year. But whether the planting platforms were raised above the original surface or represent remnants of that surface left between the canals, whether dams controlled water levels or not, whether planting surfaces seasonally flooded or not, what crops had been grown, how nutrients had been cycled, how much construction and function varied in time and space, and nearly every other issue remains, to some degree, under investigation. In contrast, the *chinampas* constitute the main extant example of intensive wetland agriculture, albeit much reduced from their sixteenth-century extent, and research has determined a model of *chinampa* construction and function as well as suggesting their sustainability over several millennia (Sluyter, 1994). Canals separate planting platforms built up out of alternating layers of lake mud and vegetation. Willow trees line the edges of the platforms to stabilize them. Platforms range around ten meters in width and one hundred meters in length. The water in the canals dampens temperature fluctuations, thus mitigating frost damage, subirrigates crops by infiltrating the platforms, and harbors fish. Mud from the canal bottoms provides fertilizer. Overall, the *chinampas* have a rectilinear plan, with ranks of rectangular fields separated by canals on which boats carry produce to market. Such features combine in a labor-intensive system to produce sustainably high yields.

INIREB thus used the *chinampas* as the model for redeployment of precolonial intensive wetland agriculture in the lowlands, with consequent negative results. During the 1970s and early 1980s, INIREB worked with farmers from the Basin of Mexico to build *chinampas* in Veracruz and Tabasco (Gómez-Pompa et al., 1982). Near the ruins of Zempoala, INIREB built an experimental field complex at its La Mancha research station in order to study biodiversity, nutrient cycling, yields, labor intensity, and other issues. The most ambitious project, however, was the Camellones Chontales of Tabasco—part experiment, part demonstration, and part CBNRM (Community Based Natural Resource Management). Backers included state and federal governments, the IAF (Inter-American Foundation), the INI (Instituto Nacional Indigenista), and the community of Chontal natives who would farm the fields. The project was so ambitious in scale that instead of manually building up fields out of alternating layers of mud and plants, INIREB employed large mechanical dredges to build sixty-five platforms, called *camellones*, each about thirty meters across and one hundred to three hundred meters long. The Chontales themselves did not participate in the construction, only in the subsequent farming.

The Camellones Chontales did not meet expectations. In 1987, Mac Chapin, then of Cultural Survival, an organization that advocates for the rights of native peoples, surveyed several INIREB projects, including the Camellones Chontales. He claimed that the *chinampa* model had "seduced" INIREB to the point of blinding it to seriously flawed assumptions (Chapin, 1988a, 1988b, 1991). A series of miscalculations seemed to support his conclusions. The *chinampa* model, while indigenous to Mexico, diffused from the highlands to the lowlands with the same discouraging results as westernization because, most basically, lowland field vestiges are so different from the *chinampas*. The former have much smaller platforms, interconnected in labyrinthine, curvilinear patterns, occur in backswamps, and their surfaces probably flooded every wet season, bringing an increment of nutrients and drowning pests and weeds. The *chinampas* have large rectangular platforms, occur in lakes, and their surfaces do not flood annually. Even discounting that the *chinampa* model has no precedent in the precolonial lowlands, the use of dredges in the construction of the Camellones Chontales inverted the soil profile, burying fertile lake mud beneath infertile clay and necessitating application of large amounts of organic matter and fertilizer to achieve acceptable yields. The dredges had also made the canal bottoms so irregular that the Chontales could not use their dragnets to fish. The crops, mainly vegetables grown on the Basin of Mexico *chinampas*, were exotic to the Chontales and oriented to market rather than subsistence. And the great diversity of lowland insects required large amounts of insecticide—only adding to the project's failings.

CONCLUSIONS

Development theory requires improvement so that it can begin to provide a basis for policies that actually improve social and environmental well-being in Mexico and other countries. Key to such improvement will be integration of culture theory and political-economy theory in a way that reveals the linkages between unworldment and undevelopment. Only then will reworldment be as possible as it is essential to the redevelopment of places undeveloped by colonialism and misdeveloped by (post)colonialism.

To reiterate and emphasize that point through illustration, even INIREB's intention to base sustainable development on a native ecology instead of on westernization went awry because of the complex, long-term process that interwove unworldment,   undevelopment,   von   Humboldt's   influence,   and   the material/conceptual transformation of landscapes—not only those of the Gulf lowlands, but those of the Basin of Mexico. As in the Gulf lowlands, (post)colonial misdevelopment of the Basin of Mexico accelerated in the twentieth century (Ezcurra 1990a, 1990b). The population of Mexico City increased from 137,000 in 1800, to 541,000 in 1900, to 13.8 million in 1980, much of that growth being due to migration from rural Mexico caused by agricultural misdevelopment. Mexico City's built-up area increased in parallel to that population growth: from 10.8 square kilometers in 1800, to 27.5 square kilometers in 1900, to 980.0 square kilometers in 1980. Drainage has virtually dried up the lakes that dominated the precolonial basin, causing the aquifer that supplies the city to fall in level by more than nine meters and become polluted by wastewater. Water shortages are chronic, with about 30% of water consumed in the basin being pumped in over the mountains from neighboring drainages. The salinized soil of the dry lakebeds, already noted by von Humboldt (1966, vol. 2, pp. 170-171) in 1803, prevents all but some poor halophytic pasture from growing; consequently, dry-season dust storms blow salt and clay particles contaminated with sewage into the city. Moreover, as the clays on which Mexico City rests dry and contract, infrastructure sinks. Even the drainage system is sinking and thus decreasing in gradient and efficiency. During the late precolonial period, some 10,000 to 20,000 hectares of *chinampas* existed, about half of them in Lake Texcoco around Tenochtitlán and the other half in Lake Xochimilco and Lake Chalco (Sluyter, 1994). By von Humboldt's visit only a small fraction survived, even fewer by the twentieth century (Armillas, 1971). The Basin of Mexico thus epitomizes (post)colonial misdevelopment.

Despite that failure, the UN (United Nations) organizations directly concerned with development, the World Bank and UNDP (United Nations Development Program), do not consider the millennia of knowledge manifest in the surviving *chinampas* to be relevant to development. Instead, in 1987 UNESCO (United Nations Educational, Scientific, and Cultural Organization) designated the "Historic Center of Mexico City and Xochimilco" as a "World Heritage Landscape" in recognition of its precolonial and colonial architecture and because the *chinampa* system, its last few hectares barely surviving in the shrunken remnant of Lake Xochimilco, "testifies to the efforts of the Aztec people to build a habitat in the midst of an unfavorable environment" (retrieved 9 October, 2003, from whc.unesco.org). The UN organizations responsible for development thus ignore the *chinampas*, except perhaps as a tourist attraction. And the UN organization responsible for culture archaeologizes the *chinampas*.

Current development efforts, dominated by political-economy theory, as is manifest in the division of labor between the UNDP and UNESCO, make the archaeologized *chinampas* unavailable for application to development within the Basin of Mexico. Yet, through that same archaeologization process, the *chinampas* become part of Mexico's romanticized Aztec heritage and thereby become available for application to development of such supposedly hostile environments as the Gulf lowlands. For INIREB, as for Sartorius before, the precolonial vestiges of intensive agriculture in the Gulf lowlands merely suggested the possibility of (post)colonial development. The actual model for such development had to come from elsewhere: for Sartorius, from Europe; for INIREB, from the ancient Aztecs—those supposed Romans of the New World.

Recent developments at FAO (United Nations Food and Agriculture Organization) suggest some hope for the integration of culture theory into development. The FAO, which has responsibility for some agricultural aspects of development, has begun to become involved in research, preservation, and applications regarding so-called GIAHS (Globally-Important Indigenous Agricultural Heritage Systems) such as intensive wetland agriculture around Lake Titicaca in the Andean highlands (retrieved 19 January, 2004, from www.fao.org). There, local Quechua and Aymara farmers have revamped more than a hundred hectares of long abandoned, intensive wetland fields (Erickson and Candler 1989; Erickson 1992). Scientists provided an initial model for field construction and function, but the local farmers have implemented and incrementally adjusted that model. The project has accomplished both of its goals: an experimental test of construction and cultivation techniques hypothesized on the basis of excavations of precolonial fields, and the production of bumper crops of (truly) postcolonial potatoes.

Only a much broader effort to integrate culture theory and political-economic theory as well as the functions of such organizations as the UNDP and UNESCO will result in better development theory and application. Such integration has the potential to explain the processes through which colonialism and (post)colonialism have unworlded precolonial worlds and apply that understanding to development that actually improves social and environmental well-being in a (truly) postcolonial world (Sluyter & Siemens, 2004b).

Department of Geography and Anthropology, Louisiana State University, Baton Rouge, Louisiana, USA, www.ga.lsu.edu

NOTES

[1] Although my views do not necessarily reflect those of the people acknowledged, I thank Anne-Marie d'Hauteserre and Theano S. Terkenli for organizing, editing, and inviting me to contribute to this volume, Ulrike Leitner for sharing insights on von Humboldt's unpublished materials, the Cartographic Section of the Department of Geography and Anthropology of the Louisiana State University for the figures, Kent Mathewson for many stimulating conversations about von Humboldt and intellectual history, and the manuscript reviewers.

[2] Throughout, quotes come from the 1966 AMS Press facsimile edition of the 1811 Longman edition of the *Political Essay* (von Humboldt 1966), translated into English from the French edition of 1811 by John Black, because that facsimile enjoys wide accessibility. I have, nonetheless, checked all passages critical to the analysis against the 1811 French edition and, as appropriate, indicate any differences of significance in brackets.

REFERENCES

Adas, M. (1989). *Machines as the measure of men: Science, technology, and ideologies of western dominance*. Ithaca: Cornell University Press.

Armillas, P. (1971). Gardens on swamps. *Science*, 174, pp. 653-661.

Blaut, J.M. (1993). *The Colonizer's model of the world: Geographical diffusionism and Eurocentric history*. New York: Guilford Press.

Brand, D.D. (1959). Humboldt's Essai Politique sur le Royaume de la Nouvelle-Espagne. In J.H. Schultze (Ed.), *Alexander von Humboldt. Studien zu seiner universalen geisteshaltung* (pp. 123-141). Berlin: Verlag Walter de Gruyter.

Chapin, M. (1988a). The seduction of models: Chinampa agriculture in Mexico. *Grassroots Development*, 12 (1), pp. 8-17.

Chapin, M. (1988b). The Seduction of models: Chinampa agriculture. *The DESFIL Newsletter*, 2 (1), 3, pp. 22-24.

Chapin, M. (1991). Travels with Eucario: In search of ecodevelopment. *Orion,* 10 (2), pp. 49-58.

Clavigero, F.J. (1780-81). *Storia antica del Messico* (Vols. 1-4). Cesena: G. Biasini.

Clavigero, F.J. (1917). *Historia antigua de México* (Vols. 1-2). J. Joaquin de Mora (Trans.) & L. Gonzalez Obregon (Ed.). Mexico: Departamento Editorial de la Dirección General de las Bellas Artes. (Original work published 1780-81)

Cortés, H. (1988). *Cartas de relación.* Mexico City: Editorial Porrúa.

Denevan, W.M. (1992). The pristine myth: The landscape of the Americas in 1492. *Annals of the Association of American Geographers,* 82, pp. 369-385.

Díaz del Castillo, B. (1986). *Historia verdadera de la conquista de la Nueva España.* Mexico City: Editorial Porrúa.

Erickson, C. (1992). Applied archaeology and rural development: Archaeology's potential contribution to the future. *Journal of the Steward Anthropology Society,* 20, pp. 1-16.

Erickson, C., & Candler, K.L. (1989). Raised fields and sustainable agriculture in the Lake Titicaca Basin of Peru. In J.O. Browder (Ed.), *Fragile lands of Latin America: Strategies for sustainable development* (pp. 230-248). Boulder: Westview Press.

Ezcurra, E. (1990a). De las chinampas a la megalopolis: El medio ambiente en la Cuenca de México. Mexico: Fondo de Cultura Económica.

Ezcurra, E. (1990b). The Basin of Mexico. In B.L. Turner II, W.C. Clark, R.W. Kates, J.F. Richards, J.T. Mathews, and W.B. Meyer (Eds.), *The earth as transformed by human action: Global and regional changes in the biosphere over the past 300 years* (pp. 577-588). Cambridge: Cambridge University Press.

Finck, H. (1871). Account of antiquities in the State of Vera Cruz, Mexico. *Smithsonian Annual Report for 1870*, pp. 373-376. Washington: Smithsonian Institution Press.

Florescano, E. (1994). *Memory, myth, and time in Mexico: From the Aztecs to independence.* Austin: University of Texas Press.

Franch, J.A. (1973). Juan de Torquemada, 1564-1624. In H.F. Cline (Ed.), *Handbook of Middle American Indians,* (Vol. 13), *Guide to ethnohistorical sources, part two* (pp. 256-275). Austin: University of Texas Press.

Glacken, C.J. (1967). *Traces on the Rhodian shore: Nature and culture in western thought from ancient times to the end of the eighteenth century.* Berkeley: University of California Press.

Gómara, F. López de. (1987). *La conquista de Mexico.* Madrid: Historia 16.

Gómez-Pompa, A., Morales, H.L., Jiménez Avila, E., & Jiménez Avila, J. (1982). Experiences in traditional hydraulic agriculture. In K. V. Flannery (Ed.), *Maya Subsistence: Studies in Memory of Dennis E. Puleston* (pp. 327-342). New York: Academic Press.

Gómez-Pompa, A. (1990). Seduction by the chinampas. *The DESFIL Newsletter,* 3 (4), 3, pp. 6-8.

Humboldt, A. von. (1810a). *Recueil d'observations astronomiques, d'opérations trigonométriques et des mesures barométriques* (Vols. 1-2). Paris: F. Schoell.

Humboldt, A. von. (1810b). *Vues des cordillères et monumens des peuples indigènes de l'Amérique.* Paris: F. Schoell.

Humboldt, A. von. (1811). *Essai politique sur le royaume de la Nouvelle-Espagne avec un atlas physique et géographique, fondé sur des observations astronomiques, des mesures trigonométriques et nivellemens barométriques* (Vols. 1-3). Paris: F. Schoell.

Humboldt, A. von. (1815). *Personal narrative of travels to the equinoctial regions of the new continent, during the Years 1799-1804.* Helen Marie Williams (Trans.). Philadelphia: M. Carey. (Original work published 1814-1831)

Humboldt, A. von. (1941). *Viaje a las regions equinocciales del nuevo continente.* L. Alvarado (Ed. and Trans.). Caracas: Biblioteca Venezolana de Cultura. (Original work published 1814-1831)

Humboldt, A. von. (1966). *Political essay on the Kingdom of New Spain* (Vols. 1-4). John Black (Ed. and Trans.). New York: AMS Press. (Original work published 1811).

Humboldt, A. von. (1970). *Tablas geográficas políticas del Reino de Nueva España y correspondencia Mexicana.* M.S. Wionczek & E. Florescano (Eds.). Mexico: Dirección General de Estadística.

Humboldt, A. von. (1980). *Cartas Americanas.* C. Minguet (Ed. and Trans.). Caracas: Biblioteca Ayacucho.

Jackson, W. (1994). *Becoming native to this place.* Lexington: University Press of Kentucky.

Leitner, U. (2000). Humboldt's works on Mexico. *HIN: Humboldt im Netz: International Review for Humboldtian Studies/Revista Internacional de Estudios Humboldtianos /Internationale Zeitschrift für Humboldt-Studien,* 1 (1), no pagination. Retrieved March 1, 2003, from www.uni-potsdam.de/u/romanistik/humboldt/hin/.

Lewis, O. (1966). The culture of poverty. *Scientific American,* 215 (4), pp. 19-25.

Livingstone, D.N. (1992). *The geographical tradition: Episodes in the history of a contested enterprise.* Oxford: Blackwell.

Lorenzana, F.A. (1980). Historia de Nueva España. Mexico: Secretaría de Hacienda y Crédito Público. (Original work published 1770)

Mayer, B. (1847). *Mexico, as it was and as it is.* Philadelphia: G. B. Fiebre.

Meadows, D.H., Meadows, D.L., Randers, J., & Behrens, W.W. (1972). *The limits to growth.* New York: Universe Books.

Mendoza Vargas, H., & Azuela Bernal, L. F. (Eds.). (2003). *Lecturas de Humboldt y México: naturaleza, cultura y sociedad.* Mexico: Universidad Nacional Autónoma de México.

Mentz, B. von. (1990). Estudio preliminar. In B. von Mentz (Ed. and Trans.), *México hacia 1850* (pp. 11-45). Mexico City: Consejo Nacional para la Cultura y las Artes.

Miranda, J. (1962). *Humboldt y México.* Mexico: Universidad Nacional Autónoma de México.

Oliver, L., Vinh H. N., & Giugale, M.M. (Eds.). (2001). *Mexico: A comprehensive development agenda for the new era.* Washington, DC: World Bank.

Pratt, M.L. (1992). *Imperial eyes: Travel writing and transculturation.* London: Routledge.

Puig-Samper, M.A., & Rebok, S. (2002). Un sabio en la Meseta: El viaje de Alejandro a España en 1799. *HIN: Humboldt im Netz: International Review for Humboldtian Studies/Revista Internacional de Estudios Humboldtianos/Internationale Zeitschrift für Humboldt-Studien,* 3 (5), no pagination. Retrieved March 1, 2003, from www.uni-potsdam.de/u/romanistik/humboldt/hin/.

Ronan, C.E. (1973). Francisco Javier Clavigero, 1731-1787. In H.F. Cline (Ed.), *Handbook of Middle American Indians,* (Vol. 13), *Guide to ethnohistorical sources, part two* (pp. 276-294). Austin: University of Texas Press.

Said, E.W. (1979). *Orientalism.* New Yor k Vintage Books.

Sartorius, C.C. (1961). *Mexico about 1850.* Stuttgart: F.A. Brockhaus. (Original work published 1858)

Siemens, A.H. (1980). Indicios de aprovechamiento agrícola prehispánico de tierras inundables en el Centro de Veracruz. *Biótica,* 5, pp. 83-92.

Siemens, A.H. (1990). *Between the summit and the sea: Central Veracruz in the nineteenth century.* Vancouver: University of British Columbia Press.

Siemens, A.H. (1998). *A favored place: San Juan River wetlands, central Veracruz, A.D. 500 to the present.* Austin: University of Texas Press.

Skerritt Gardner, D. (1993). Colonización y modernización del campo en el centro de Veracruz (siglo XIX). *Cuadernos de Historia,* 2 (5), pp. 39-57.

Sluyter, A. (1994). Intensive wetland agriculture in Mesoamerica: Space, time, and form. *Annals of the Association of American Geographers,* 84, pp. 557-584.

Sluyter, A. (1997). On excavating and burying epistemologies. *Annals of the Association of American Geographers,* 87, pp. 700-702.

Sluyter, A. (1999). The making of the myth in postcolonial development: Material-conceptual landscape transformation in sixteenth-century Veracruz. *Annals of the Association of American Geographers,* 89, pp. 377-401.

Sluyter, A. (2001). Colonialism and landscape in the Americas: Material/conceptual transformations and continuing consequences. *Annals of the Association of American Geographers,* 91, pp. 410-428.

Sluyter, A. (2002). *Colonialism and landscape: Postcolonial theory and applications.* New York: Rowman & Littlefield.

Sluyter, A. (2004). Alexander von Humboldt, (Post)colonial Mexican Landscapes, and Economic Development as a Cultural Phenomenon. The Association of American Geographers, 100th Annual Meeting, Philadelphia, Pennsylvania, March 14-19. Abstract in *Abstracts of the 100th Annual Meeting of the Association of American Geographers,* CD-ROM, no pagination.

Sluyter, A., & Siemens, A.H. (Guest eds.). (2004a). *Special issue: Native food production knowledge systems and practices. Agriculture and Human Values,* 21, pp. 101-261.

Sluyter, A., & Siemens, A.H. (2004b). Native food production knowledge systems and practices: Alternative values and outcomes. *Agriculture and Human Values,* 21, 101-103.

Strebel, H. (1883). Die ruinen von Cempoallan im Staate Veracruz (Mexico) und mitteilungen über die Totonaken der jetztzeit. Hamburg: H. Strebel.

Torquemada, J. de. (1969). Monarquía Indiana (Vols. 1-3). Mexico City: Editorial Porrúa. (Original work published 1615)

Turner, B. L. II, & Butzer, K.W. (1992). The Columbian encounter and land-use change. *Environment,* 34 (8), pp.16-20, 37-44.

Wolf, E.R. (1982). *Europe and the people without history.* Berkeley: University of California Press.

KATRIINA SOINI, HANNES PALANG, KADRI SEMM

FROM PLACES TO NON-PLACES?

Landscape and sense of place in the Finnish and Estonian countrysides

Setu village in its everyday contrasts: laid back and vital, source Helen Soovali

INTRODUCTION

This chapter studies rural landscapes in Finland and Estonia, in the outer rim of Europe. Peripheries are thought to be backward, but this backwardness carries traditions and establishes a lifeworld for local people. At the same time, global change has its influence even in the peripheries. For Estonia and Finland, this means a decline in rural population, the abandonment of rural taskscapes[1] and the need to finding ways to keep the countryside alive through new, or alternative, ways of life. This in turn means changes in culture, landscape, places and place-making.

117

T.S. Terkenli & A-M. d'Hauteserre (eds.), Landscapes of a New Cultural Economy of Space, 117-148.

Landscapes of insiders are changing into landscapes of both insiders and outsiders, and in some places the insideness may disappear altogether.

In the following we will explore whether and how the processes of unworldment and deworldment (as explained by Terkenli, 2004) take place in the Finnish and Estonian countryside. The paper will concentrate on places-in-memory, virtualization of places and creation of non-authentic places thereby seen from the perspective of the insiders. The background of this interpretation will be the process of rural landscape transformation currently going on in Finland and Estonia. Extensive areas formerly booming with rural life are now under economic, social and cultural reconstruction, leaving an imprint on both the physical and the social landscape. At the same time, rural landscape and life are becoming increasingly fashionable, people settle in the countryside not to work in agriculture, but rather to perform the lifestyle.

In this chapter we approach the rural landscape as 'a place' for different actions, functions and actors with various views. We start with a brief conceptual and theoretical introduction into the implications and transformation of the sense of place in the context of rural landscapes. Then we examine two processes concerning rural landscape management. The first one deals with the different views of landscape, the dichotomy of insideness and outsideness, which is present in the national programs for conserving cultural landscapes and in everyday life at the local level. Another process is about segmentation of places happening in the rural communities in both countries: how some places are thinning out and even disappearing on the one hand, and how new places emerge and are established on the other. We argue that new places emerge on top of the older ones, as meanings of the place are being lost or transformed. The knowledge about places so far has been gained in stable societies. Estonia and Finland have not had such a stable history. As far as we know, there has been no study about how places are being carried through formation changes. In this paper we try to compare the changes in a stable society and a changing society. So the question here is how big a change in a place can be so that we still have the same place and not a new one. Does qualitative change create a new place? Although these processes within rural areas – valuation and segmentation – have many advantages, they may lead to the loss of sense of place, to the creation of non-places and to placelessness as well as to the uneven development of rural areas.

RURAL LANDSCAPE AS PLACE

Place and sense of place

Landscape is perhaps one of the most contested and misunderstood concepts in (human) geography and other disciplines. Different theoretical approaches and methodologies have been created for exploring and explaining the relationship between humans, environment and landscape (Cosgrove, 1998; Jones, 2003). Humanistic geographers have drawn on the concept of place when studying the human/environment or human/landscape relationship. Contrary to landscape, place provides a midway between objective fact and subjective feeling, which Entrikin (1991) calls the betweenness of places. It appears as a combination of objects and meanings that differs somehow from its surroundings, regardless of scale.

Place is a central word of our everyday language, although there exist differences in the ways it is used in various languages. In addition to its everyday usage, the concept has been adopted by various academic disciplines. Place, though, like landscape (Cresswell, 2003; Setten, 2004), has remained elusive and difficult to deal with in research and despite various attempts, place does not lend itself to a single definite interpretation. Its definitions and approaches vary from phenomenological ones to more or less behavioral ones. Lewis (1979, p. 28) has stated that "It is often easier to see its results in human behavior than to define it in precise terms".

Sense of place is the capacity to recognize and respond to diverse identities of places (Relph, 1976). It examines people's ties and attachment to their places, or what some (Agnew, 1987) have called the structure of feeling. The sense of place may be positive or negative, neither purely subjective nor intersubjective, and never stable (Butz and Eyles, 1997; Williams and Stewart, 1998). Sense of place is not a uniformly shared and undifferentiated concept. As Sack (1997) has noted, place constrains and enables the content and extent of our awareness and actions and vice versa. Different positions relative to places have been clarified, for example, by making a distinction between an insider's and an outsider's sense of place (Relph, 1976; Shamai, 1991). The insider's sense of place seeks to grasp the distinctive properties of particular places in terms of their meanings for those who live and work in them. An outsider looks at a place mostly in terms of objective qualities.

Relph (1976, pp. 49-55) has characterized different kinds of insiders from existential to vicarious and of outsiders from incidental to existential. At first glance, the various levels may seem artificial. However, there is not necessarily a sharp distinction between the levels, but rather a continuum from not having any sense of place to a deep commitment towards a place. The ranking procedure may assist in dealing with the different perceptions and interests evoked by landscape and in avoiding confusion between the different scales and perspectives that are increasingly present in rural landscapes. Considering the dichotomy of insider and outsider, the main repercussion from a rural landscape point of view is that the insiders are often unable to realize the values and meanings of the landscape they belong to. Furthermore, they tend to use different language and concepts from outsiders when communicating about their everyday lifeworld. For these reasons, the insiders are in danger of remaining 'outsiders' in rural landscape planning and management.

Both *unworldment* and *deworldment* transform the dimensions of place, and this chapter will examine how these processes are reflected in place and sense of place. It has been suggested that place has shifted from the conventional notion of place as coherent, bounded and stable to a diffused space of flows (Castells, 1996). On the other hand, the value of located place has been rediscovered (Massey, 1995; Agnew, 2003) and the sense of place even intensified rather than diminished (Pringle, 2003). Furthermore, Terkenli (2005) suggests that along with a process of deworldment *a new collective sense of place* at the global level has emerged. This sense of place is widely appreciable and consumable by virtually everyone around the world.

Augé (1995) has introduced *non-places* as places of supermodernity, which refers to the excess of time, space and individualism: History becomes current events, space becomes images and the individual merely a gaze. Non-place is "the others' space without the others in it, space constituted in spectacle, a spectacle itself already hemmed in by the words and stereotypes that comment upon it in advance in the conventional language of folklore, the picturesque or erudition" (Augé, 1998, p. 106). Thus, a clear distinction between non-place and place is that the former lacks any chronological connection to a broader physical, cultural or emotional content, unlike the latter. Airports, motorways and hotel rooms are typically planned to be non-places, whereas rural places have represented and provided an opposite for them. However, as rural places including nature, culture and history are increasingly considered as resources and basis for economic activities, it might be useful to be aware of the possible emergence of non-places in rural space.

Scale and time

When places are considered, there is the ever-present issue of scale and time (Smith, 1993; Terkenli, 2005). Scale is the criterion of difference not so much *between* places as between *different kinds* of places. Scale both contains social activity and, at the same time, provides an already partitioned geography within which social activity takes place (Smith, 1993). There is a difference between the public symbolic sense of place that is connected to larger areal units from city to nation and the field of sense of place that applies to small areal units. In a way, both types are present simultaneously due to the intersubjective, social and cultural character of the sense of place. Separating distinct dimensions of social spatiality in studies has caused scale to collapse into an over-generalized chaotic concept, as Terkenli (2005) has noted.

Time is involved both in landscape and place in various ways. In the early 1980s, Cosgrove (1984, 1998 p. xiv) argued that landscape history should be understood as part of wider history of economy and society[2]. So every socio-economic formation tries to create its own landscape, by wiping off the land the uses and symbolic values of previous formations and replacing these with its own. A formation should here be understood as a set of political, economic, social and cultural conditions prevailing in a society.

In Western Europe, change from one formation to another has been gradual, and transitions (such as from feudalism to capitalism) took decades, if not centuries to reach completion. Each formation also had time to develop its own landscapes. A political organization defines land use patterns that reflect the legal system of the country (Olwig, 2002; Mitchell, 2003). Through arts and communication, a landscape ideal is created, and that later becomes the yardstick for policy and tourism. It contains memories of the past (so vividly described by Schama, 1995) and preconditions for the future.

The recent political changes in Eastern Europe have shown that changes in socio-economic formations create time barriers in the landscape that are not transparent – younger people who have not lived in the previous formations are unable to understand those landscapes (Palang et al., 2002, 2004a, 2004b). Both the historical and the everyday contexts are thus significant for landscape study. Changes in landscape and their understanding by locals are not as sharp and sudden as political changes. There is a certain time lag, as Maandi (2003) has argued: the physical traces linger and the stories are passed on to the following generation even if the practices have ceased.

As we have seen, places are in transition both globally and locally in various ways. At the same time and as a part of this process, traditional way of living in rural areas in which the sense of place has been very constitutive, is nowadays seriously challenged. How do these processes meet in rural Estonia and Finland?

RURAL LANDSCAPE DEVELOPMENT IN ESTONIA AND FINLAND – DIFFERENCES AND SIMILARITIES

Estonia and Finland has many ethnic and linguistic similarities (that make the two countries thus comparable). It is not possible to compare them and we are not aiming at comparing them. But the similarities make the study fascinating. They are both situated in the European periphery and there are at least three sets of processes that have affected the cultural landscape of rural areas during the past two centuries in

both countries: national awakening, modernization of agriculture and institutionaliza-
tion of the setting of rural areas, including land reforms related to various policies.

By the beginning of the 17th century, both countries were part of the Swedish
kingdom. Their major difference by that time was that in Estonia feudalism had been
introduced by the German-speaking estate-owners, while Finnish peasants enjoyed
personal freedom under the Swedish-speaking upper class. Later, the Swedish
kingdom had to turn both countries over to the emerging Russian Empire – Estonia
was annexed in 1710 and Finland a century later in 1809. For more than a century
thereafter their history was rather similar.

Land parceling was one of the most important landmarks in the land use history
of the two countries. In Finland, the general parceling of land started in 1757 and in
Estonia in 1816/19. The main outcome of this reform from the rural landscape
structure point of view is the dispersion of previous villages, which established a
community in a physical and 'governmental' sense (Troska, 1987; Katajamäki,
1991; Hyyryläinen, 1994; Saarenheimo, 2003).

National awakening started in Finland in the first half of the 19th century.
National and especially rural landscapes played an important role in strengthening
national identity in the arts (Häyrynen, 1997; Björn, 2003; Sooväli et al., 2003).
Finland became independent in December 1917, Estonia followed in February 1918.
However, Estonia was annexed by the USSR in the summer of 1940 and regained its
independence only in August 1991.

This divergence translated into four distinctive time layers in the landscape in
Estonia. We can speak about the *estate landscapes* of the parcellation time ended by
the 1919 land reform, nationalizing all lands belonging to non-Estonians; the *private
farm landscapes,* following parcellation and independence; the *collective farm
landscapes,* resulting from collectivization in 1949-50 and representing the Soviet
value system; and, finally, *the post-modern landscapes* of the 1990s. The latter are
characterized by nostalgia for the golden 1930s and by the desperate need to find
new identities under the new political, economic and cultural conditions. The
tendency is now for people to leave the countryside for towns, so that large
agricultural areas are no longer used. The postmodern landscapes of today lack both
ideological and economic continuity and feature a multitude of absent landowners
and diverse land use practices (Palang et al., 2004b).

Rural areas of Finland developed rapidly since late 19th century and especially
during 1920-1950. The number of independent farms increased, as crofts became
independent in 1918. In addition, after WWII, 425,000 Karelian people were
resettled[3] and about 30,000 farms and other residences with a piece of land were
established for them. Many new agricultural and social innovations were introduced
in the rural areas in the first half of the century. However, by the 1960s, the
agricultural boom period was over and overproduction became a problem

(Katajamäki, 1991). In a way, this development peaked after Finland joined the European Union in 1995. Since then every fourth farm has been closed, while the average size of the farms has increased. Finnish agriculture has specialized and concentrated both at the local and at the regional scale. Farmers used to comprise the majority of the village population at the beginning of the 20th century, but are a minority now. Most people work outside their home villages. In addition, there is an increasing number of part-time residents, who are looking for their place and role in the rural communities.

Since the 1990s, rural landscape management has become an issue in planning and management on the national, regional and local level, in both countries. Previously, landscape development resulted directly from other policies. Now, landscape management is an integral part of agricultural, environmental and rural policies, which are grounded in the multifunctional character of agriculture. Besides its other values, landscape has increasingly been considered as a source of livelihood for local people through tourism and rural product markets.

Various programs and procedures related to landscape management are being implemented in both countries. The Finnish Agri-Environmental Programme includes measures for landscape management, and a similar programme is being launched in Estonia. Both in Estonia and Finland, national, regional and local programs exist for identifying valuable landscapes and for creating procedures in order to conserve and maintain these landscapes (Alumäe et al., 2003; Palang et al., 2004c for Estonia, Ministry of Environment 1993 for Finland). The question, however, remains as to how these initiatives affect the sense of place and the future of places.

RURAL PLACES THROUGH TIME

The following section shows the ways in which rural places in both countries struggle for their continuation. We demonstrate this struggle through three stories, one from Finland and two from Estonia. The countries shared a similar history for most of the past 200 years, with the exception of 50 years from 1940 till 1991. However, the processes taking place are surprisingly similar – some places cease to exist, some places retain their meanings and structures through constant cultural redefinition, some places obtain new meanings. However, the cases we present here are only cases, it does not mean that in Estonia all places wither out or commercialize and in Finland they transform.

The Väälma story – a place that has vanished

Palang and Paal (2002) describe Väälma as a small farm in the village of Metsaküla in the eastern part of the Kõrvemaa forests in the north of Estonia. The area is boggy with several low islands of dry soil, sand and gravel reaching out of the layers of peat. A long esker stretches east of the area. Bogs and mires are covered with forests extending some 6-7 kilometers eastward and tens of kilometers in other directions. It is located about 6 km off the nearest main road.

When the story starts, there was a forest. One day, people appeared who cut down the forest, turned it into field and pasture and sorted out their relations with the landlord in order to keep the place. Then, their descendants created the place that is remembered today as a safe home. Finally the people left or died, the fields were abandoned and grasslands became overgrown, the place names and legends are being forgotten. And after some decades it will be a forest again.

Usually, when delimiting places, one has in mind spatial boundaries or locations. This story shows that places have limits in time as well.

Acquiring the place

One way of place emerging is through place-naming. The more "local" people get, the more exact the names become. A folk legend tells:

> A bear came down the hill and settled down on an oat-field (Kaerassaare). There he stayed, time and time again crossing the end of a river (Jõeotsa) and a sandy area (Liivaku) to go hunting. Once, he quarreled (Riiussaare) with another bear, but they could not sort it out. So they went to a wise man (Targa) to judge the quarrel. The first bear was found guilty; he got angry and died of plague (Katku).

This legend illustrates how the separate farms in the Metsaküla village acquired their names (Figure 1). However, two of the farms, Väälma and Tuimõisa, do not figure in it, because they are considered to be older than the legend (Tambek, 1983). The origin of the name Väälma is somewhat unclear. Varep (1985) has found the name Väälma (*im Wehlmegschen*) in official documents written in 1693. Tambek (1983) tried to reconstruct it as *Väälma – Vealma < vee all maa* (land under water). More probably the name is derived from the family name Wälman < wäl man 'good man' in Swedish.

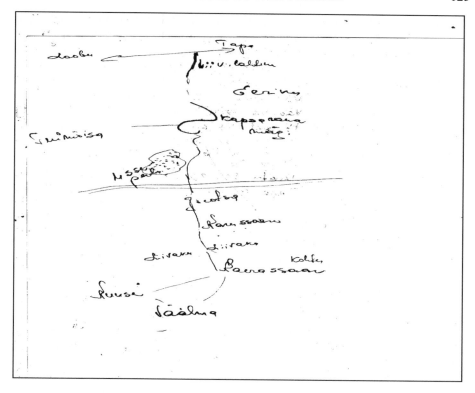

Figure 1. The road to Väälma, passing most of the farms, as remembered by one of the descendants of the last owners, source: author

According to written history, the Faehlmann - Wälman family has lived in the parish of Kadrina at least since 1630, when the forefather Väälma Madis was mentioned in an agricultural census by Swedish authorities. According to oral history (family history), after the Great Nordic War (1700-1721) the landlord gave a piece of land to Väälma Madis and his descendants, because he was hiding a Swedish soldier in his household. And hence also the family name.

As this kind of free-holding was very extraordinary in those times, the lifestyle of that family differed from their neighbors'. The feeling of freedom has supported that kind of specific family-centered worldview where outsiders have never been too welcome. The house, situated in the middle of swamps and forests, was like a nest to the family where everybody could return when times were unsettled or when there was danger. The house has had a central place in the family life and it is no wonder that people still dream of Väälma, even though there is nothing left of it nowadays.

All oral stories describe Väälma as a free farm and the people as being more than just peasants. All written data, on the contrary, show Väälma as a farm located on a separate piece of land, providing statute labor for the landowner and paying rent. As the amount of arable land was not very large and therefore the rent low, the tenants were able to develop a smithy that remained a characteristic of the farm for many decades. Being a smith raised the social status of the farmer and provided money to be saved, so that in 1865, the new master, Madis, was able to buy the farm for property.

Thus 1865 marks the end of the estate landscape in this particular spot. The dream of freedom had come true; the ancient rights, once granted by the King of Sweden and later robbed by local German land owners, were restored. Madis and his sons were now in control of their land, which could only increase their pride. This was also expressed in the desire to give their children as good an education as possible. The farm was still busy with smith-work that provided income, but later Madis (1810-1881) had to divide the farm between his two sons.

Creating the place

This was the story of acquiring the place. Finally, after all sorts of controversies the official registers showed the people living there as owners. The time for creating had come. This creation is directly tied with Bertha (more often called Mamma), granddaughter of Madis the farm-buyer. But before Bertha with her husband and children appeared on the scene, several important changes had to take place in Väälma.

The second son of Madis, Juhan (1851-1939), inherited the "new" part of the farm called Uue-Väälma that was 33 hectares in size. He, although having received a good education, continued traditional smith-work. In 1878-80, he erected most of the buildings that will later feature in our story. He married Marie Worteil and they had 3 daughters. As education was also considered a sign of freedom, the daughters Bertha and Elsa got an excellent education by the standards of those times.

In 1904, Bertha married Anton Lindam (known as Papa). The couple lived elsewhere, but spent summers in Väälma. In 1916, they left for Kazakhstan to escape mobilization for WWI, only to return in October 1921 with 4 children and plenty of ideas on how to develop the farm. *"The buildings were tumbling down, the smithy was devastated, grandfather was ill, people were afraid for their lives"*. Now starts the story told by Bertha's grandchildren about the golden home with a granny a little bit selfish and strict, but still providing a safe haven when necessary.

Bertha's mother died in 1929, her father followed ten years later. All fieldwork was done by family members, there have never been farm hands, the story says. Rye went well, Bertha was thrifty, even avaricious, and made money out of eggs, butter, milk - they used to have 4-5 cows. While farming had been the main income for Bertha's parents, the new generation got theirs from other sources. Smith-work also provided income. They took timber from their own forest and, in 1929, erected a new house. Within 15 years Bertha and Anton managed to build up the farm so that, by 1939, they had bought the former estate in Kõnnu, opened a dairy there and a shop. This is the Väälma that is now remembered in the stories of Bertha's grandchildren. In all these stories, the "soul" of the place is Bertha. She is the one who seemed to be in command; places are remembered according to what Momma had done there. She, on one hand, pushed her children out (letting them get married and settle down in other places), on the other hand, she always provided a safe haven for those in trouble.

Losing the place

The latter became especially evident in the 1940s. War rolled over the country and those family members in trouble found a refuge here. Some fled from mobilization in the German Army, some from Soviet power, some living in towns were afraid of air raids or other military events. Väälma was a refuge.

Restoration of Soviet power in 1944 indicated the start of the decline. Land had been nationalized already in 1940, although the owners were given a state guarantee allowing them to use the land "forever". That guarantee lasted only till 1949, when collectivization started. Anton lost his shop, his shares in the inn, all his property. Finally, in 1948, Anton got a gun and ended his life. Repressions followed Bertha's son-in-law and two of his own sons fled to Sweden in 1944. Son Endel who had participated in the Finnish War since 1941, returned from the Finnish army and got repressed as well.

From 1948 on, Bertha managed the farm on her own, with occasional help from Endel – he lived only some 5 kilometers away in Mõndavere. Instead of the formerly crowded place full of life, there remained only Bertha, now 66 years old, and her daughter Aino (whose husband Leo had been deported to Siberia in 1950) with her three small children. Finally, in 1951, Bertha joined the kolkhoz. The kolkhoz cultivated the lands till the end of the 1950s, but, as it was a remote and not very fertile place, the fields gradually turned into grasslands and in the early 1960s, alien calves destroyed Bertha's small garden.

In 1964, Bertha finally gave in, took her goat and moved to Aino's place in Kadrina. The kolkhoz and later the Lahemaa National Park mowed the grasslands.

The house stood empty till 1973, when Bertha's son Konstantin took the house apart and built it up again in Rakvere. Bertha herself died in Rakvere, in her house, in 1981. That was the physical end of Väälma. Still, as often as possible, some of the family members went to visit the place. Now it lives only in memories and stories. The narratives describe first the demolition of the buildings, later the abandonment of the whole area. And when the last family members die, also the memories will be gone.

The Setu case – transformations through time

Setumaa, together with the island of Kihnu, are the only two areas in Estonia that have retained their distinctive folk culture and the area attracts a lot of tourism attention. There also seems to be disagreement about the future of this distinct cultural landscape.

Changes through time

Semm and Palang (2004) and Semm and Sooväli (2004) have shown how changing power relations have influenced the Setu landscape. These studies, based on interviews with local people and on text analysis, show how Setumaa has been perceived by its inhabitants, throughout the 20th century.

Land played an important role in the life ways of the 1920-30s. Land served as a source of income; rituals, customs and beliefs were connected to it. People were eager to carry on their customs; the communal system functioned, as people communicated within villages and identified themselves by villages. Interestingly enough, contemporary scientific literature and media depicted Setumaa (together with Saaremaa) as the least-developed parts of Estonia, which could be used for studying ancient life styles (Sooväli et al 2004).

During Soviet occupation, traditional culture was not valued. Identity was preserved thanks to the older generations, who kept the traditions. This happened in suppressed and hidden ways, while socialist life patterns overshadowed them with their symbols. By the early 1990s, people were no longer familiar with historic life ways and culture. In the traditional landscape, meanings were connected to spiritualized nature and religious elements; in today's cultural landscape these symbols are important as preservers of heritage and thus of the (local) identity. Local festivals aimed at remembering and reproducing traditions carry a meaning of reunification for the Setus.

Nature also plays an important role – land was important in the 1920s, and berry forests were named as one of the most valuable landscapes in 2000 (Alumäe et al., 2003). Often people see Setumaa as a tourism area with beautiful nature. Past life ways and their imprint on the cultural landscape faded from memory during Soviet times. Is the reason for revitalizing these meanings today mainly connected with recreation purposes? In the present Setumaa, activities include walking in the forest, picking berries and mushrooms, but no longer cultivating fields.

People try to create an ideal mental image of the area. Here, one can draw parallels to Cosgrove's (1984, 1998) ideas about fixating the iconographic landscape. For example, tourism booklets show traditionally characteristic building styles, women in their folk costumes singing or pictures of cultivated rural fields (Semm and Sooväli, 2004). At the same time, Olwig (2004) has shown how this representation of landscape creates a circular reference between the representation and the "real" life. Once a landscape gets depicted in a tourism brochure or described in a travel guidebook, those visiting will be looking for that very image, and they are disappointed if they do not find it there. This might mean a decrease in income. To avoid that, the "real" landscape is being re-arranged to match the representation, to show the tourists what they want to see. This might lead to seasonal effects (tourism-oriented show-time plus tourist-free real-time) and a sort of schizophrenic misunderstandings of what is "real" and what is not.

What could happen to the cultural landscape in Setumaa is that Setumaa will enter that trap, being viewed as a seasonal summer area, where old lifestyles are forgotten or preserved only in archives or in traditional symbols in the landscape, which do not possess any meaning for the local people. Now traditions are again becoming significant in today's community. The more active people introduce the old traditions or symbols of Setumaa both to the locals and to the general public. Runnel (2002) describes how local authorities are trying to coddle that feeling of communal society, valued in the first place only by old local people. Local culture tends to be influenced by mass media. The role of local community as a creator of identity is weakening. As Annist (2004) concludes, village life is now sustained by tourists, which has turned rural life into a strange disrupted, fragmented, seasonal event. However, a scenario study conducted to show the future options for the landscapes of Setumaa (Palang et al. 2000) demonstrated that locals welcomed both continuation of the more traditional agricultural lifestyles, the counter-landscape of the Soviet times, and the tourism landscape, modernized and symbolizing Scandinavian welfare to them.

This leads to a discussion about the role of different actors in the landscape, or about whether landscape should be seen as practice (for local people with their meanings and everyday actions) or as scenery (for tourists to gaze at). This dilemma may be seen as a power play between different interest groups (Alumäe et al., 2003; Palang et al., 2004c). At the same time, there are different ideas about what is authentic and which time should serve as a baseline for defining authenticity (Gustavsson and Peterson, 2003).

Different interest groups – different places?

In Setumaa, one can bring out three interest groups in the landscape. First, the local Setu people, who constitute the majority population in Setu areas, live their everyday life, concern themselves with everyday problems and do not bother with the aspirations of the Setu activists. Maybe this is due to the heritage of passivism of the Soviet time, or due to their own understanding of their Setu identity. Being less receptive to innovations, they are the ones who maintain the characteristic ways of life. Because of high unemployment, they cultivate land, pasture animals and raise income from picking mushrooms and berries. They go to church and carry on their customs. Annist (2004) calls this group "everyday-locals" – they keep the authenticity of culture and could be considered the existential insiders (or dwellers) in Relph's (1986) sense.

Secondly, there are the local Setu activists and intellectuals of Setu origin living elsewhere, who try to maintain the traditions erased during the Soviet time. This group has an important effect on today's landscape. They are concerned about retaining the heritage, both for the local residents and for the wider public as well. They often live in memories rather than the actual situation and think they are in a position to judge what is authentic and what is not and thereby able to define traditions.

These two groups understand the landscape so differently that it is very difficult to find a common ground between them. Preserving culture cannot rely only on external symbols. These symbols should be understood and given meanings. This links back to the discussion about places in the beginning of this chapter – place is more personal than landscape and places could be given meanings only through personal experience, from inside. This is reflected in today's cultural landscape – young people in Setumaa get involved in place-making by learning local history and taking part in the (re)introduction of customs by external agents.

The third group includes people who are interested in the landscape of Setumaa as tourists – they treat the landscape as scenery. Since the fall of the Soviet system an increasing number of summerhouses has been set up and instead of permanent residence people visit Setumaa more in summer. The meanings they give to the surrounding environment differ greatly from the local ones. This group brings economic revenues to the region. For them the landscape symbols could be *tsässons*, museums displaying both objects of nature and cultural events. This important possibility raises Setu people's self-assurance and prevents understanding the Setumaa landscape as just a beautiful summer landscape.

Annist (2004) underlines that cultural or historical identity may not always be the local identity. It is often an identity constructed from outside, which might be detrimental to the local one. Local people take the advice of the "authenticity watchdogs" too seriously and this might result in locking the local life in the past. The locals are expected to pay too much attention to culture. When the past shadows the present and everyday local practices, it becomes difficult to define and maintain the local identity. The past should be a step towards the future, not the focus of life. And everyday practice should maintain the traditions.

The Setu cultural landscape is at crossroads. On one hand, the distinctiveness of the cultural landscape, the visibility of time layers, the continuation of taskscapes that are considered traditional, attract visitors. And these visitors come to see the authentic, "real past", thus launching the circular referencing process Olwig (2004) refers to. At the same time, this "real past" evolves, as change is always part of the landscape. Local people are the ones who, paradoxically, carry out the change while representing the past for the visitors. Discussion in geography about whether landscape should be handled as scenery or practice is thus mirrored in real life.

The Mommila case: Are there any signs of new cultural economy in an ordinary Finnish village?

Mommila represents a typical Finnish village of rural Finland, in terms of socio-economic structure, changes in agriculture and as a site of recreation[4]. Yet, its history, which is strongly related to the estate of Mommila and the landscape around Lake Mommilanjärvi, provides added value for studying the influence of global trends in the transformation of place at the local level. This transformation is described here through the insights of the insiders of the village – farmers, other rural residents and part-time residents.

The history of Mommila village

Mommila village is located in the Southern part of Finland, in Häme County, and in the southernmost part of the Lammi municipality. The landscape is dominated by a large, plain field area surrounded by forests. An old estate with many aesthetically and historically valuable buildings adds special value to the landscape. Lake Mommilanjärvi, which dominates the map of the village, is mostly hidden from the passers-by. The lake is rather shallow and subject to eutrophication, but, as a result of water conservation actions, the state of the lake is stable. It has a recreational meaning especially for the summerhouse settlement on the western side of it (Figure 2). The southernmost part of the lake belongs to the neighboring municipality, Hausjärvi.

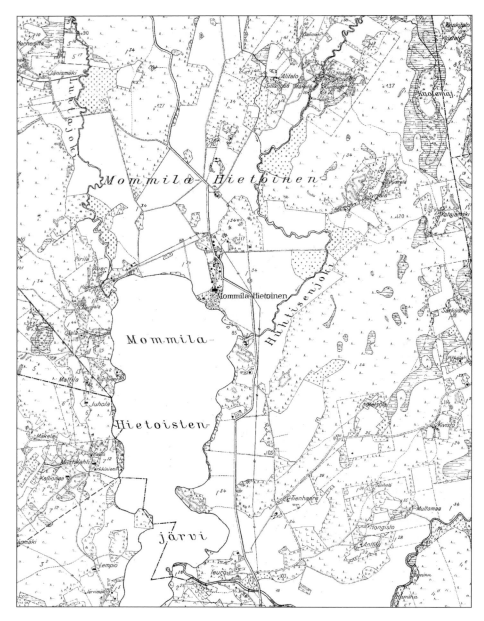

Figure 2. The map of Mommila is drawn in 1933, completed in 1938. after that many new farms have been established in the western and northwestern side of the village and the summerhouse settlement has emerget on the eastern side of the Lake, source: author

The history of Mommila is dominated by the Mommila estate. The origin of the estate dates back to the 13th and 14th centuries. At the beginning of the 20th century, during the time of councilor Kordelin (1903 – 1917), it gained meaning in the Finnish agricultural and political history. Kordelin introduced many agricultural and social innovations in the village, built a school, completed the building of the church and created economic activities ranging from agriculture to small-scale manufacture. The estate and the village used to be self-sufficient and traded in food and other products with Helsinki and St. Petersburg. The size of the estate at its largest was 3,000 hectares and it employed almost 500 people (Pänkäläinen, 2001). Although Kordelin was committed to developing agriculture and the local community, he wanted to keep his distance from the employees. He never won their confidence and favor (Koskue, 2002).

Kordelin was murdered in 1917 nearby the village in a riot related to WWI and the civil war. After that, the estate changed hands many times and lost much of its land and wealth. This historical trajectory was also related to the changed position of the crofts of the estate: in Finland the crofts became independent in 1920. In 1927, the estate was bought by a family from outside the village, who still owns it. In the 1940s, the estate was divided into three parts, all located at the eastern side of the lake and physically separated from the rest of the village by a large field area and the main road crossing the village.

After WWII, one third of the land of the Mommila estate was taken for the new farms established for the Karelian evacuees and others, who needed to be resettled after the war. Almost one hundred small farms were organized with some hectares of cultivated land and a piece of forest. Thus various cultures and ways of life were suddenly brought together, which changed the physical, social and cultural structure of the landscape of Mommila[5] (Koskue, 2002).

At present, about 300 people live in Mommila. Their main source of livelihood is agriculture, although over half of the working population earns a living outside the home. There are a number of summerhouses, thanks to the relatively close location and good traffic connections to the growing cities. In summertime, the population of the village doubles. The population of the village thus constitutes a mixture of different social backgrounds: previous crofters, successors of the estate, those of Karelian extraction, urban summerhouse residents, some of whose are previous dwellers of the village. Numerous recreational activities take place in the village, such as hunting and other sports, where a village committee organizes events and creates services. Their distant location from the centre of the Lammi municipality has brought the people together. For instance, they have successfully fought for maintaining the village school and for implementing road reconstruction.

As in many other villages in Finland, a village plan was drawn by local and part-time residents in 1997. The plan maps the possibilities, weaknesses, strengths and threats to the village, including the landscape, nature and heritage of Kordelin as resources for tourism and other business activities. In addition, a landscape management plan was drawn by the Rural Advisory Centre in 2003 for every village of the Lammi municipality, including Mommila. The plan comprised of landscape analysis, with proposals for improving and highlighting naturally and culturally valuable sites and views.

In the following we will explore the implications of the recent rural development on the sense of place and possible emergence of new cultural economy of space at local community level.

Farmers – between place and economy

Farmers are usually the existential insiders of the local community. Typically, a farm is simultaneously the farmers' home and working place, and most of the farmers have lived on their farm for their entire life. In Finland, continuity from generation to generation, a central value for farmers, has persisted until today (Kumpulainen, 1999; Silvasti, 2001). However, the general trends of agricultural development in Finland as described above are evident also in Mommila. Many of the small farms, former crofts and farms established after the war have closed. The rest have leased the fields and try to adapt to new requirements and regulations introduced by agricultural and environmental policy as well as to diversify their livelihood in order to be able to keep both farming practices and the place itself.

Although agricultural policy is increasingly highlighting environmentally sound farming practices and the idea of multifunctional agriculture, the productionist ethos is still alive among the farmers. This is revealed, for example, in the farmers' aesthetic appreciation of the landscape: for them the most pleasant scenery is comprised of well maintained fields, vital crops that are clean of weeds and large field shapes. The new environmentally-oriented land use forms such as buffer strips in the field margins and butches in the middle of the fields, may disturb the 'good farmer's gaze', (see also Burton 2004), even if the farmers admit that those areas might be useful for wildlife or make the landscape more diverse. Appropriateness regulates the farming practices within the compass of agricultural and environmental policy. For the farmers, the previous taskscapes may become chronologically more organized landscape, stringing people and events out, suggesting separation and discontinuities. Landscape and biodiversity management are considered as important, but belonging primarily to those, who are interested in game management or who have tourism as a main livelihood.

Besides continuity, land ownership has been another central value for the Finnish farmers (Silvasti, 2001). Previously, it was easier for anyone living in a community to name the fields and their owner(s), as the fields were located around the farms. Now, almost 40% of all the fields are leased (Finnish Agriculture, 2004) and, consequently, a farmer's fields may be dispersed throughout the community. Furthermore, it is typical that a farm is comprised of many 'side-farms': the fields of which are considered as productive means rather than as places with names and identity. The relationship between farmer and the land, which has traditionally been very close, is gradually acquiring new forms and shaping the farmers' sense of place.

In some cases, farmers have been obliged to give up farming due to ageing or economic reasons, sell the cattle and fields and turn the active farm into a living-place. Along this process, the farmers' gaze that has previously been tightly focused on the growing crops, extends to a larger scale; to the history, ecological processes and landscape of the village. Farmers are really unable to perceive the 'values' of their environment in the broader sense, so long as they are actively farming. When they give up farming and related daily tasks, they are able to see landscape development and the history of the village more clearly and they may find and create new places in their everyday environment – they view the landscape more like an outsider. According to Ingold (1993, 2000), the previous 'taskscapes' become chronologically organized landscapes, stringing people and events out, thus suggesting separation and discontinuities.

Sense of place has carried the farmers over troubles for centuries. Currently, due to rapid changes in policy and ownership, the farmers have to get used to giving up places and getting attached to new ones. Buying, selling and leasing of land have always been a part of rural history. This time, however, the identity of the farmers is also being challenged; the farmers are asked to adopt new roles, for example as environmental managers. According to Robinson (1991), a place is significant for social identity. When identities are modified, so too are places. Similarly, it is possible to argue that identities have to be reformulated along with changing places. Now, places of landscape are gradually changing, but the identity of the farmers changes much more slowly resulting in contradictory feelings, which were expressed by a farmer in the following way:

> F: If they pay enough, I will sow this field full with willow (laughing) ... I am such a materialist...To be serious, I would not close that landscape [which was closest to the farm and visible from the yard], but a distant piece of field, I would certainly do that, if I were paid enough.

He considers new possibilities to keep on farming, but, at the same time, desires to maintain his most meaningful place of landscape, home, out of the business, as private and personal as if in order to protect his identity as a farmer. .

Part-time residents safeguarding their paradise

On the other side, there are the part-time residents, who are temporarily involved in the community. Some of the part-timers have had a long relationship to the village; they have memories from childhood or are former residents of the village. Others have become residents of the village accidentally or after a long, deliberate search.

A common feature for the part-time residents is their wish to create a place of their own, a paradise. A summer cottage is a place, which provides an alternative to life in the city, representing many, nostalgia, peace, nature and wilderness. Those who had spent their childhood in Mommila wanted to transfer similar experiences to their children. Place provides such a sense of continuity for them.

In addition to the symbolic values, the local community and landscape provide many instrumental values for them: goods, services. Landscape and nature are assessed on the basis of the needs and desires of the part-time residents, for example the forests as a source of berries and mushrooms and the lake as a site for fishing and swimming, thus recreational activities. The part-time residents do not necessarily want to become active members of the community; they would rather keep their distance. Still, they wish that the landscape and life in the community would provide a viable and healthy frame for their paradise preserved into the future.

Compared with the farmers, part-time residents look at the landscape from a different time perspective. For them, landscape is more like a continuum, which they drop in from time to time, rather than a process or taskscape. For these reasons, stability creates feeling of safety and continuity, and sudden changes in their everyday landscape, for example the cuttings of the forests in the neighbourhood of the part-time residents, are judged more severely.

Non-agrarian residents - mediators of transformation

The non-agrarian residents, the biggest group of the three, occupy the middle position in several ways between the farmers and the part-time residents. On one hand, compared to the farmers, they have not encountered immediate changes in their everyday places of the landscape. They live their everyday life around their homeyard in Mommila and perform their activities elsewhere. On the other hand, as they are more often "inside" the landscape than the part-time residents, landscape changes and challenges are not that visible for them.

At first glance, it seems that the non-agrarian residents have restricted possibilities for place-making, because they do not own farm/land. In addition, their social relations are usually more oriented outside the village. For these reasons, the non-agrarian residents may be more able to see the aesthetic, historical and natural values of the landscape and the possibilities these provide for local livelihood. In addition, they would be more able to adapt and put into practice new innovations related to the rural tourism as their identity is not necessarily bound with the physical places. Thus, they may play a role as interpreters of the values and mediators between the other interest groups inside and outside the community.

From one place to several places - place segmentation

As Sack (1997, pp. 7-11) describes, in simple societies places are less differentiated than in the modern society. Places are domains for many activities from instrumental to ritual and each place is thick with meaning. In the modern world, people are linked globally and human actions are interrelated at many levels. Places tend to be more differentiated and specialized, which has led to a segmentation of places and identities. The division between the private and the public space has become clearer, as people tend to protect their home and privacy. But, at the same time the local people are more and more aware of and influenced by the outsiders' gaze and in that way, the borderlines between the private/public and insider/outsider have blurred. Along this process landscape or a part of it may be detached from the place identity of the insiders, a part of which it has once been.

At the beginning of the 20th century, Mommila used to provide places for both work and leisure. Nowadays, Mommila represents different kinds of places, aims and services for different interest groups. For some people, Mommila is a place for economic activities and hobbies, while for some others the village is just a place for living or simply for rest.

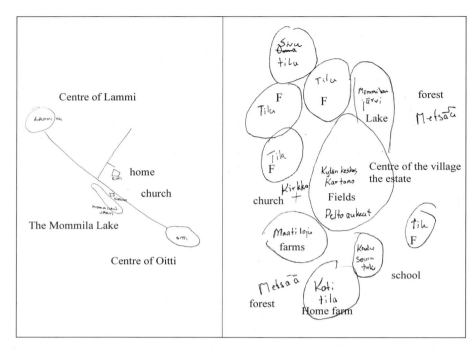

Figure 3. Everyday landscapes and places drawn by a farmer and his wife living in Mommila village. The wife is working outside home, as the husband in farming and making wood products at home, source: author

There are, however, some sites, in which interests are overlapping constituting "thick places". The church and the graveyard make people conscious of the history of the village and create a sense of belonging. The community house regained its meaning through voluntary and collective action during a renovation project, and the village shop is a symbol of persistence and vitality of the village. These places "belong" to everyone.

In some cases, meanings are actively maintained and even newly created. Lake Mommilanjärvi is an almost invisible place and accessible only from a few locations. Every year, since 1991, there is a rowing event organized by the locals together with locals of the villages of the neighboring municipalities[6]. The event emerged by an initiative a part-time resident and has become an ever growing and versatile summer-event with thousands of visitors. The aim of the event is to finance the restoration of the lake and to make people aware of the state of the lake and, in this way, commit the people to its conservation. This event brings the lake and its significance to the attention of both the local people and the tourists.

There are also places that are mentally distant or even "empty" of meanings. The lake, large field areas and a road divide Mommila physically, but also socially into various parts: the estate with its surroundings, summerhouse settlement and the rest of the village, which has previously belonged to the estate. People blame and try to hide the social barriers, but, at the same time, they like to adhere to them, as local hierarchies and social barriers maintain a link to the history of the village and in a way, strengthen the identity of each group.

Despite of a small decline, Mommila has managed to maintain the population, unlike many other rural villages in Finland. Local people are satisfied or even proud that almost all houses of the village are still inhabited, at least temporarily. Empty houses are, however, not necessarily empty of meanings. They are places of memories, as Väälma for a while, but in the process of time they easily become symbols of village decline. The inhabited houses with the smoke coming out of chimney are not considered to be the only signs of life in the landscape. People, who remember olden times, were missing people from the landscape; people working on the field and walking along the road and having time for a word.

The future of the village seen from inside and outside

Landscape perceptions are compared with memories and future expectations. Longing for the past might imply dissatisfaction with the present or fear for the future. Both permanent and part-time residents are worried about the future of the village: the future of farming, the ageing of the population, and even the growing number of unmarried farmers. Farmers' life is considered to be too stressful and, in addition to the farmers themselves, many other people in the village worry about their health. In this situation, people like to turn to the past and start to remember the time of hay poles and milk platforms, which symbolizes the good old and more peaceful days. In this way, the sense of place easily becomes nostalgic (Butz and Eyles, 1997). People are linked to the history of the village with various kinds of bonds, but insecure feelings about the future are shared collectively.

Several plans have been drawn for the village, some of them by the locals themselves and the others by environmental authorities or experts of regional planning or landscape management. Although these planning processes were thought to be participatory, the plans were usually guided by outsiders – and typically the local people themselves start to think like an outsider during the process – and probably for that reason, most of the ideas related to the commercialization of local values are still unexploited.

The study also pointed out that the local people are not very well aware of all the planning activities going on, while the landowners, especially, are primarily interested in the outcome that is related only to their own land. However, people are gradually becoming aware of values and expectations set by the outsiders on their everyday environment. One interviewee wondered why they were chosen for the landscape study, even though they did not belong to the 'cultural landscape area'. This points out how the valuation process itself may differentiate places and landscapes in the minds of the local people, and suggests the need to pay attention to the fact that the intervention, which touches on the sense of place, should be carried out at the local level.

DISCUSSION AND CONCLUSION

In this chapter we have explored the transformation of places through three case studies from the periphery of Europe, in Finland and Estonia. The case studies show how changing socio-economic formations have influenced landscape changes in both countries. Estonia has a multitude of time layers in its landscape, each of which has had a drastic influence. After Estonia regained its independence processes have been twofold. On one hand, they have brought along marginalization and outflow of both population and as economic activities. On the other hand, there is a growing interest towards revitalizing traditions. In Finland, changes have not been as abrupt as in Estonia and historical continuity in the landscape is much more evident, although the situation is different related to the marginal regions and cultures in Finland, e.g. in Lapland and in Archipelago, which have not been dealt with in this chapter. Still, places are gradually being transformed at various scales, from an unconscious to a conscious stage and gradually more and more towards commodification. It links directly to the process of deworldment, as defined by Terkenli (2005): the society of spectacle with a new collective sense of place based on transcending geographical barriers of distance and of place, articulated at the global scale, widely palatable and consumable by virtually everyone around the world.

Breaks in time and scale – emergence of non-places?

Unlike non-places, places typically have a physical and chronological connection to the broader physical, cultural or emotional context. Our case studies pointed out that some places gradually wither away in time, as the Väälma and Mommila cases show – places cease to exist as physical reality. On one hand, all traditions (in terms of value systems, symbolism, land tenure) have been interrupted by the sequence of changing socio-economic formations, on the other hand, this uprootedness has caused cultural resistance, a (re)valuing of landscapes reminiscent of the past and greater awareness of the value lost. These cases have also provided an excellent chance to examine which features survive the formation changes and which are the most volatile ones. Surprisingly, as Palang et al (2004d) argue, the primary power structures are among the most resistant ones to all sorts of changes - collective farm centers are located at the same places occupied a thousand years earlier by local chiefs.

The Setu case study has pointed out that, as a result of changing power relations during the Soviet time, the rural landscape has become a "foreign country" for younger generations. People born in the early eighties and later do not understand rural life, for them it is totally foreign. They are unable to understand, how places functioned, and which values were involved in those landscapes. For them, collective landscapes are remnants of the past, something whose existence is over. Thus they become outsiders and strive towards building something new on the ruins of previous formations.

In areas where collective farms were not that well off, the sense of place is still present, although marginalization and connected processes take place – abandonment, negligence etc. Finally, these places end up in memory, as the Väälma story illustrates. They cease to exist physically, they remain in the memories of those who once had some sort of connection with the place, but, as time passes, people die and the fate of these places could be compared to that of American or Australian ghost towns.

But as Mommila shows, the increase in social and cultural diversity lets us argue that this has helped the community to adapt to the changing situations, insist on and reproduce local culture and resist the disappearance of the barriers of distance and of place, the deworldment. Consequently there is no space for the emergence of non-places at the moment.

Towards place-making – deworldment of places

Whereas some places are lost in one way or another, new places are emerging. In some places, as in Setu, people try to revitalize the past and to convert it into something new, usually into places that could be used by the tourism industry. This signals the creation of new stories and legends that better fit the ideas or expectations of the tourists – see the discussion on circular reference above. The threat here is museumisation of the places, freezing of the past or creating a pseudo-past.

Then, there are places like Mommila, where there exist possibilities for active place-making, but the process has not yet started. People have so far been able to earn their livelihood otherwise and are locked in thinking that these places exist primarily for themselves. Only a few locals are considering the ways they could make business out of providing products and services for the part time residents, and some locals have become aware that landscape could be a product. Finally, there exist many places or villages in both countries, which have 'no values' from the outsiders' point of view. These landscapes do not correspond to the images the outsiders have of rural landscape. In addition, these places are not necessarily accessible or included in landscape managements programs. Still, they establish a lifeworld, a place with meanings for local people.

In the process of place-making, the different views of outsiders and insiders meet and the question of power becomes critical. Is it the place of the local people, who are in a position to take care of their lived environment and living culture as it is expressed in their landscapes? Or is it about outside control, by policies, different economic and public awareness campaigns through which outsiders seize control over the landscape traditionally controlled by insiders? The Setu case shows that insiders are willing to develop their culture in their own way. The outsiders are also increasingly willing to develop the rural culture according to the values and desires of themselves. They see versatile needs and possibilities to conserve the rural idyll. There have already been voices asking for protection of the "rural acoustic heritage" (Carles et al., 1999).

There is enormous pressure from outside to fit this development into some sort of set of rules or regulations, as described in textbooks of environmental management, using legislative, fiscal and public information instruments. There are policy measures that support certain types of activities without consideration of how they fit into a specific situation, legal measures forbidding some and allowing other activities, as well as planning and other procedures that aim at doing the best, but sometimes achieve another level of standardization. Many of these programs except the legal ones leave on the paper and the policy makers are wondering how to make local people adapt and implement them. One of the reasons is that some researchers (e.g. Burton, 2004) have found out that adaptation also requires redefinition of symbolic meanings and that takes time. We may also ask, what we may lose if the local people, for example the farmers, start really redefining their basic values. Thus, there is always the power struggle between those for whom landscape is practice and those for whom landscape is gaze (Setten 2003 provides an interesting insight into this theme). One party desires to live in that landscape, maintain one's places, but raise income from it; another wants to see the landscape as it once was, be that an example of the 1930s or the 1980s, but nonetheless, the landscape has to be there and meet expectations.

Considering power relations inherent to place-making, one should keep in mind some characteristics of rural communities. Although global processes are ever present, rural communities are also 'located places', in which face-to-face contacts occur and are needed. In addition, borderlines and rights concerning private and public spheres are part of tacit knowledge. Place-making requires crossing borderlines, both physical and social, and creating new kinds of shared experiences of the lifeworld as well. What could these shared meanings be? Are they shared virtually or in real life? Another question is, whether it is possible to find "real" "authentic" and "shared" meanings at local level as the life spheres have become segmented.

Authenticity and the new cultural economy of space

Change has always been part of the landscape in both countries. In Estonia, the landscapes consist of several time layers, each of which has had different social and cultural settings. In Finland, changes have been smoother, but still always present. As soon as we start talking about "authenticity", we have to define the time line. We can speak about what was authentic for a certain time period or we may create the illusion of authenticity (see Gustavsson and Peterson, 2003; Dovey, 1989; Sooväli, 2004; Korhonen et al., 2003 etc.). The time line is different for insiders and outsiders: The insiders of the place are about continuation, humility, appropriateness, memories, today and tomorrow. Outsiders are about control, conservation and museumisation – usually lacking both a physical and temporal connectivity between and to the places.

The case studies have shown that there is no such thing as a "traditional" landscape or "traditional" place. "Authentic" places are created either unself-consciously or self-consciously in harmony with the context, implying a pre-condition of meaning. Appropriateness and meaning have been central place values, especially to existential insiders, for example to the farmers. For this reason, it may be assumed that construction of "inauthentic" places might require a change in the social and cultural norms. That takes time and compels us to commit ourselves to the social and cultural sustainability.

However, as a result of processes of the new cultural economy of space, this process is not as slow as expected. Landscape management programs, which might be regarded as a sign of the new cultural economy of space, may finally result in the emergence of museums of local lore, with the hope that tourists will come and bring money. A part of the tradition, formerly communicated orally from one generation to another, is thus transformed into easily-consumable culture wrappings. This, in turn, creates another layer in the landscape, the dichotomy between places meant for insider consumption and non-places meant for outsiders. For them, an authentic illusion of places is being staged and landscape managed so that it looks beautiful, however not necessarily bearing the functions it used to. Although it is easy to criticize landscape management interventions about the standardization of places and landscapes and encouraging local people to make "inauthentic" places, they may arouse an interest in taking a closer look at one's landscape, place and identity in new ways, as if for the first time.

Local processes have always been influenced by the outer world and all innovations reach even the most remote periphery. Rural life driven by agriculture is being replaced by a mixture of work-leisure-tourism; local functions, tasks and local knowledge are in the process of change, becoming easily-consumable and much less personal. Urry (1995) has noted that the various demands in the countryside may 'leave no stone unturned' by the twenty-first century. He is suspicious if there is much remain of the mystery, memory or surprise left in the countryside, because such places may well have by the been consumed, used up. Furthermore, concerning the segmentation, Sack (1997) argues that places can become so thin that everything is virtually alike, significant differences disappear and little is to be gained by looking elsewhere. So far, these kinds of scenarios may seem doubtful, as neither Finland nor Estonia is among the global top tourism destinations, although they are gaining popularity at the national level. Still the processes of unworldment and deworldment should not totally be ignored. They may rapidly become manifested also in the rural periphery touching the geographical selves of both the insiders and outsiders.

ACKNOWLEDGEMENT

Parts of the research for this chapter have been supported by the Estonian Science Foundation grant No 5858 and by the Finnish Cultural Foundation, Häme Regional Found.

Katriina Soini, MTT/Environmental Research, Jokioinen, Finland
Hannes Palang, Institute of Ecology, Tallinn Pedagogical University and Institute of Geography, University of Tartu, Estonia
Kadri Semm, Estonian Institute of Humanities, Tallinn. Formerly Institute of Geography, University of Tartu, Estonia 10

NOTES

[1] With the term 'task' Ingold refers to any practical operation, carried out by a skilled agent in an environment, as a part of his or her normal business of life. The tasks are constitutive acts of dwelling, not suspended in a vacuum. A taskscape exists only as long as people are actually engaged in the activities of dwelling.

[2] Having said this, the reader should also take into account Cosgrove's own introduction to the 1998 edition of his book, Olwig's (2002) comments on landscape as scenery, and Cosgrove's (2003) idea of the two discourses in landscape.

[3] Karelians were settled almost all over Finland, except the majority of Lapland, Swedish-speaking municipalities and some areas on the north-western coast (Laitinen 1995, pp. 111-112).

[4] Granberg (2004) for example has studied the changes from agricultural to tourism in some more marginal regions with a specific cultural characteristics, such as Pyhätunturi, holiday resort in Lapland, a small family enterprise on a rural island in the southeastern Finnish archipelago and the municipality of Suomussalmi, which is located at the Finnish-Russian border and carries memories of the battles of the Finnish Winter War.

[5] Topelius (1875) has vividly described the differences in the character of the people of the Finnish counties. According to his typology, the Karelian people are hospitable, extrovert and vivacious, while the Häme people are described as silent, slow and hardworking. Sallinen-Gimpl (1994) has studied the encountering of the two cultures, the Karelian and the host one in a rural community. She found differences between the two cultures in the structuring of time, perception and experience of social and private space, group and family allegiance and loyalty, and relation to the environment (including the natural environment, food, work techniques)

[6] The event itself takes place on the other side of the border of the village and the municipality.

REFERENCES

Agnew, J. 1987. *Place and Politics: the geographical mediation of state and society*. Boston and London: Allen and Unwin.

Agnew, J. 2003. Classics in human geography revisited. *Progress in Human Geography* 27, 5 pp. 605-614.

Alumäe H., A. Printsmann and H. Palang 2003. Cultural and Historical Values in Landscape Planning: Locals' Perceptions. In H. Palang, G. Fry (eds.) *Landscape Interfaces: Cultural heritage in changing landscapes*. Boston and Dordrecht: Kluwer Academic Publishers, pp.125-146.

Annist, A. 2004. Maaelu pärast põllumajandust. *Kunst.ee* 1, pp. 71-74

Antrop, M. 2000. Background concepts for integrated landscape analysis. *Agriculture, Ecosystems and Environment*, 77, pp. 17–28

Augé M. 1995. *Non-Places. Introduction to an Anthropology of Supermodernity*. Verso, London and New York.

Augé M. 1998. *A Sense for the Other. The Timeliness and Relevance of Anthropology*. Stanford University, California.

Burton, R., J., F. 2004. *Seeing Through the 'Good Farmer's' Eyes: Towards Developing an Understanding of the Social Symbolic Value of 'Productivist' Behaviour*. Sociologia Ruralis 44(2): 195-215.

Butz, D. and Eyles, J. 1997. Reconceptualizing Senses of Place: Social relations, ideology and ecology. *Geografiska Annaler* 79B, pp. 1-25.

Björn, I. 2003. Muuttuva maalaismaisema. In V. Rasila, E. Jutikkala, A. Mäkelä-Alitalo (eds.) *Suomen maatalouden historia I. Perinteisen maatalouden aika. Esihistoriasta 1870-luvulle*. Helsinki. Suomalaisen Kirjallisuuden Seura. pp. 598-619.

Carles, J.L., Barrio, I.L., de Lucio, J.V. 1999. Sound influence on landscape values. *Landscape and Urban Planning*, 43 (4), pp. 191-200.

Castells, M. 1996. *The Rise of Network Society*. Oxford and Massachusetts: Blackwell Publishers Ltd.

Cosgrove, D. 1984, 1998. *Social Formation and Symbolic Landscape*. Madison, Wisconsin: University of Wisconsin Press.

Cosgrove, D. 1998. Cultural Landscapes. In T. Unwin (Ed.), *A European Geography* (pp. 65-81). London: Longman

Cosgrove, D. 2003. Landscape: ecology and semiosis. In H. Palang, G. Fry (eds), *Landscape Interfaces. Cultural Heritage in Changing Landscapes*. Boston and Dordrecht: Kluwer Academic Publishers, pp. 15-21.

Cosgrove, D. 2005. Los Angeles and the Italian *città diffusa*: landscapes of the cultural space economy. *This volume*.

Cresswell, T. 2003. Landscape and the Obliteration for Practice. In K. Anderson, M. Domosh, S. Pile and N. Thrift (eds.) *Handbook of Cultural Geography*. Sage Publications, pp. 269- 282.

Dovey, K.1989. The quest for authenticity and the replication of environmental meaning. In Seamon, D. & Murgerauer, R. *Dwelling, place and environment*. Oxford and New York: Columbia University Press Morningside Edition, pp. 33-50.

Entrikin, J.N. 1991. *The Betweenness of Place: Towards a Geography of Modernity*. The Johns Hopkins University Press, Baltimore, MD. *Finnish Agriculture* 2004. Agrifood Research Centre, Economic Research, Helsinki.

Granberg, L. 2004. From Agriculture to Tourism: Constructing New Relations Between Rural Nature and Culture in Lithuania and Finland. In Alanen, I (eds.). *Mapping the rural problem in the Baltic countryside: Transition processes in the rural areas of Estonia, Latvia and Lithuania*, Aldershot and Burlington, Ashgate Publishing Ltd, pp. 159-179.

Gustavsson, R. and Peterson, A. 2003. Authenticity in landscape conservation and management – the importance of the local context. In H. Palang, G. Fry (eds), *Landscape Interfaces. Cultural Heritage in Changing Landscapes*. Boston and Dordrecht: Kluwer Academic Publishers, pp. 319–356.

Häyrynen, M. 1998. Isänmaan äidinkasvot. The images of fatherland. In M. Luostarinen & A. Yli-Viikari (eds.) *Maaseudun kulttuurimaisemat. Rural landscapes in Finland*. Helsinki. Finnish Environment Institute. pp. 30-34.

Hyyryläinen, T. 1994. *Toiminnan aika. Tutkimus suomalaisesta kylätoiminnasta*. Vammala. 222 p.

Ingold, T. 2000. *The Perception of the Environment*. London: Routledge.

Jones, M. 2003. The Concept of Cultural Landscape – Discourse and Narratives. In H. Palang, G. Fry (eds), *Landscape Interfaces. Cultural Heritage in Changing Landscapes*. Boston and Dordrecht: Kluwer Academic Publishers, pp. 21-52.

Katajamäki, H. 1991. Suomen maaseudun suuri kertomus. The Long and Widing Road of Rural Finland. *Terra* 103, 3, pp. 173-183.

Koskue, K. 2000. *Lammin pitäjän historia. Osa 3: vuodet 1917-1995*. The History of Lammi. Part 3, pp. 1917-1995. Jyväskylä, Lammin kunta.

Korhonen, T., Ruotsala, H. and Uusitalo, E.(eds.) 2003. Making and Breaking the borders: ethnographical interpretations, presentations and representations. Helsinki: Finnish Literature Society.

Kumpulainen, M. 1999. *Maan ja talouden välissä. Viisi kertomusta suomalaisen maatilan ja luontosuhteen muutoksesta*. Joensuu, Joensuun yliopisto.

Laitinen, E. (eds.) 1995. Rintamalta raivioille. Sodanjälkeinen asutustoiminta 50 vuotta. Jyväskylä, Atena Kustannus Oyv.

Lewis, P. 1979. Defining a sense of place. In W.P. Prenshaw and J.O. Mc Kee (eds.) *Sense of place*: Mississippi. University of Mississippi. Jacson, MI.

Maandi, P. 2003. *Landscape and land restitution in Estonia. Geographical patterns of national history and individual memory*. Presentation at the conference Landscape, Law and Justice, Oslo, pp. 15–19 June

Massey, D. 1995. The conceptualization of space. In D. Massey and P. Jess (eds.), *A Place in the World*. Milton Keynes, The Open University.

Ministry of Environment 1993. *Arvokkaat maisema-alueet: maisema-aluetyöryhmän mietintö II. Valuable landscapes: paper* Helsinki: Ympäristöministeriö. Mietintö nro 66/1992, p. 204.

Mitchell D. 2003. Cultural Landscapes: Just Landscapes or Landscapes of Justice? *Progress in Human Geography*, 27, pp. 787-796.

Olwig, K.R. 2002. *Landscape, Nature and the Body Politic: From Britain's Renaissance to America's New World*. Madison: University of Wisconsin Press

Olwig, K.R. 2004. "This is not a landscape". Circulating reference and land shaping. In H. Palang, H. Sooväli, M. Antrop, G. Setten (eds), *European Rural Landscapes: Persistence and Change in a Globalising Environment*. Boston and Dordrecht: Kluwer Academic Publishers, pp. 41–66.

Palang, H., P. Paal 2002: Places Gained and Lost. In V. Sarapik, K. Tüür, M. Laanemets (eds) *Koht ja Paik II Place and Location. Eesti Kunstiakadeemia Toimetised, pp. 93-111.*

Palang, H., H. Alumäe, Ü. Mander 2000: Holistic Aspects in Landscape Development: a scenario approach. *Landscape and Urban Planning*, p. 50, pp. 85-94.

Palang, H., Külvik, M., Printsmann, A., Kaur, E. & Alumäe, H. 2002. Maastik, sidusus, identiteet. In Ü. Kaevats (ed), *Usaldus. Vastutus. Sidusus. Eesti sotsiaalteadlaste aastakonverents*. Tallinn: Tallinna Tehnikaülikooli Kirjastus, pp. 27–32.

Palang H., A. Printsmann, É. Konkoly Gyuró, M. Urbanc, E. Skowronek, W. Woloszyn 2004a: The forgotten rural landscapes of Central and Eastern Europe. *Landscape Ecology, in press*

Palang H., H. Sooväli, A. Printsmann, T. Peil, V. Lang, M. Konsa, M. Külvik, H. Alumäe, K. Sepp 2004b: Estonian cultural landscapes: persistence and change. In J.-M. Punning (ed), *Estonia 9. Geographical studies.* Tallinn: Akadeemia Kirjastus, pp. 155-170.

Palang, H., H. Alumäe, A. Printsmann and K. Sepp 2004c: Landscape values and context in planning: an Estonian model. In J. Brandt, H. Vejre (eds.), *Multifunctional landscapes – vol. 1: theory, values and history.* Southampton: Wessex Institute of Technology Press, pp. 219-233.

Palang H., A. Printsmann, M. Konsa, V. Lang 2004d: Ideology and tradition in landscape change. A case of Helme, Estonia. In T. Peil, M. Jones (eds). *Landscape Law and Justice* Norwegian Academy of Sciences.

Pänkäläinen, M. 2001. *Lammin pitäjän historia. Osa 2: vuode 1808-1917.* Jyväskylä, Lammin kunta.

Pringle, D.G. 2003. Classics in human geography revisited. *Progress in Human Geography* 27, 5, pp. 605-614.

Relph, E. 1976/86. *Place and placelessness.* Research in planning and design series. London: Pion

Robinson, D.J. 1991. The language and significance of place in Latin America. In Agnew, J.A. & Duncan, J.S. (eds.) *The Power of Place. Bringing together geographical and sociological imaginations.* Boston, Unwin Hyman. pp. 157-184.

Runnel, P. 2002. *Traditsiooniline kultuur setude enesemääratluses 1990ndatel aastatel.* Tartu Ülikooli etnoloogia õppetool. Tartu, Tartu Ülikool, pp. 155

Saarenheimo, J. 2003. Isojako. In V. Rasila, E. Jutikkala, A. Mäkelä-Alitalo (eds.) *Suomen maatalouden historia I. Perinteisen maatalouden aika. Esihistoriasta 1870-luvulle.* Helsinki. Suomalaisen Kirjallisuuden Seura. pp. 349-362.

Sack, R. D. 1997. *Homo Geographicus: A Framework for Action, Awareness, and Moral Concern.* Baltimore: Johns Hopkins University Press

Sallinen – Gimpl, P. 1994. Siirtokarjalainen identiteetti ja kulttuurien kohtaaminen. Cultural identity and cultural clash: the resettled Karelians in Finland. Helsinki. Kansatieteellinen arikisto 40, Suomen muinaismuistoyhdistys.

Schama, S. 1995. *Landscape and Memory.* New York: Alfred A. Knopf.

Semm K., Palang H. 2004. Lifeways in the Setu cultural landscape. *Pro Ethnologia,* in press.

Semm, K., Sooväli, H. 2004: Võimuideoloogiate peegeldused Setumaa kultuurmaastikes. In *Setumaa kogumik 2. Uurimusi Setumaa arheoloogiast, geograafiast, rahvakultuurist ja ajaloost.* Tallinn: Ajaloo instituut, MTÜ Arheoloogia keskus, pp. 19-32.

Setten, G. 2003. Landscapes of gaze and practice. Norsk Geografisk Tidsskrift, p. 57, pp.134-144.

Setten, G. 2004. Naming and Claiming Discourse. In H. Palang, H. Sooväli, M. Antrop, G. Setten (eds), *European Rural Landscapes: Persistence and Change in a Globalising Environment.* Boston and Dordrecht: Kluwer Academic Publishers, pp. 67-82.

Shamai, S. 1991. Sense of place: an Empirical Measurement. *Geoforum* 22 (3), pp. 347-358-

Silvasti, T. 2001. *Talonpojan elämä: tutkimus elämäntapaa jäsentävistä kulttuurisista malleista.* Helsinki, Suomen Kirjallisuuden Seura.

Smith, N. 1993. Homeless/global: Scaling places. In Bird, J. et al. (eds.). *Mapping of the Futures. Local Cultures, Global Change.*pp. 87-119.

Sooväli, H. 2004. Saaremaa waltz. Landscape imagery of Saaremaa island in the 20th century. PhD thesis. Tartu University Press.

Sooväli, H., Palang, H., Külvik, M. 2003. The role of rural landscapes in shaping Estonian national identity. T. Unwin, T. Spek (ed.) *European Landscapes: From Mountain to Sea. Proceedings of the 19th session of the Permanent European Conference for the Study of the Rural Landscape at London and Aberystwyth.* Huma, Tallinn, pp. 114-121.

Sooväli H., K. Semm, H. Palang 2004: Coping with landscapes — the Setu and Saaremaa people, Estonia. In: T. Peil, M. Jones (eds). *Landscape, Law and Justice* Norwegian Academy of Sciences.

Tambek, A. 1983. *Väälma talust läbi aegade ja üldjoontes Metsakülast.* Manuscript at the Estonian National Museum.

Terkenli Th. S. 2005. New landscape spatialities: the changing scales of function and symbolism. *Landscape and Urban Planning*, p.70, pp. 165-176.

Topelius, Z. 1875/1983. *Boken om vårt land*. Helsingfors, Borgå, Jockas, WSOY.

Troska, G. 1987. *Eesti külad XIX sajandil*. Estonian Academy of Sciences, Institute of History, Tallinn.

Urry, J. 1995. Consuming Places. London and New York: Routledge.

Varep, E. 1985. Ajaloolis-geograafilisi märkmeid asustuse kujunemisest Lahemaa Rahvuspargi lõunaosas. In: I. Etverk (comp.) *Lahemaa uurimused II*. Valgus, Tallinn, lk., pp. 5-20

Williams, D.R. and Stewart, S.I. 1998. Sense of Place. An Elusive Concept That Is Finindig a Home in Ecosystem Management. *Journal of Forestry*, pp. 18-24.

ANNE-MARIE d'HAUTESERRE

LANDSCAPES OF THE TROPICS:

Tourism and the new cultural economy in the third world.

Little – frequented hybrid non – heterotopia, source A.-M. d'Hauteserre

INTRODUCTION

This chapter presents a different approach to the production of tourist landscapes and destinations in the third world. Two parallel production systems, the production of space and the production of symbols have led to the creation of coherent spatial representations or narratives (Zukin, 1995) of a new symbolic economy of space; the recent cultural turn in the social sciences has led to the idea of cultural economy (Crang, 1997), which recognises that economic practices are culturally constructed. Global circuits of capital of this new cultural economy have carved tourism destinations in the third world out of their contexts. They have been constructed as

T.S. Terkenli & A-M. d'Hauteserre (eds.), Landscapes of a New Cultural Economy of Space, 149-169.

heterotopias (Foucault), i.e. as extraterritorial spaces where different behaviours from those of the larger territory in which they are enclosed are practised and in which tourists have a position of economic privilege. Tourism is but one of the circuits of capital and as a consumer of space its reorganised 'unique' serialised landscapes promise an unrivalled experience marketed in the West to seduce potential travellers.

The chapter shows how many of these tourism landscapes are products of unworldment and deworldment processes by examining how they have been (and are) actively constructed by both investors and tourists of the first world. The new cultural economy, pioneered by Southern California, has led to a 'new urbanism' and 'fantasy landscapes' (see Cosgrove, chapter 3) that it has extended to the third world where it has commodified and materialised 'difference' to render it more familiar to its consumers. Landscapes, which the flows of people, capital, and information have impressed, are here conceptualised as socially constructed records of the cultural layerings and the sedimented discourses of past and present generations, not of those who inhabit them but of those who continue to extract profits from them without having to negotiate their acceptance. The question then arises as to how these discourses are developed, made known and apprehended, who is entitled to invoke them to inscribe those spaces with meaning and what this meaning might be.

In tourism, temporal and spatial continuities have been abolished to allow the staging of physical and emotional experiences by tourists at the expense of (indigenous) historical and biographical expression. Such sites have become the expression of the imposition of a hegemonic culture with the power to naturalize and normalise such a process. Tourist destinations and their landscapes are imaginary creations – they can thus be developed as a 'place on the margin' or on the periphery of the 'real world' (hence the attraction of islands). Their celebration reveals, according to Harvey a "reactionary politics of an aestheticized spatiality" (1989). Their aestheticisation and their commercialisation belong to deworldment trends although their serialised standardisation as 'familiar' destinations for westerners indicates processes of unworldment (Terkenli, 2002; introduction).

Following Lefebvre's argument (1991) that the material order of space is a social one, I characterise these landscapes as 'quasi-objects' which Latour (1993) defines as phenomena created by humans, a social production, "the consequence of a collective human transformation of nature" (Cosgrove, 1984, p. 13). These landscapes are grounded in specific places which are not home to the tourists but which they are (feel) compelled to occupy. These quasi-objects, or 'worlds of exhibitions' (Mitchell, 2001), that contain both human and other-than human elements are spatial and visual products of colonial discourse, capital expansionism and an extension of the field of western 'adventure'. They have been developed as utopian fantasies through spatial appropriation and transformation. The local residents find themselves being-out-of-place, their space transformed into a stage for the tourist-subjects to produce their own pleasure. A few sites only are allowed to develop as 'post traditional' or non-heterotopic destinations. Our question then is to what end(s) these tourism landscapes of the Third World are constructed and whether any contestation (resistance) is possible?

TROPICAL LANDSCAPES AS A PRODUCT OF CULTURE

Who inhabits tourism destinations?

A dialogical relationship develops between landscape on the one hand and cultural organisation and belief on the other. Landscape, according to Malpas, is "that in which ... 'being in the world' is grounded" (1999, p. 15). The imposition of a particular system of meanings (or culture) always serves particular ends and particular interests and becomes manifest as built structures in the land. Landscape denotes an intimate bonding between place and culture in a dynamic and relational fashion; a landscape provides the over-arching frame, a material condition of possibility, within which experience is possible, to paraphrase Malpas (1999). Landscapes are also constructed by economic practices (of 'an invasive capitalist economy') that enable the reproduction of social life. Both cultural production and economic practice involve power. And yet, landscape displays a deceptive appearance of naturalness and transparency. For insiders, according to Cosgrove (1984, p. 18) the landscape is integrated and inclusive of their life's occurrences. Are tourists integrated in the landscape? Should they be considered as outsiders or insiders?

Where are the 'hosts'? One wonders where the displaced communities have gone. To the Zurich zoo which displays in a 10,000sqm greenhouse a piece of Madagascar rainforest? The periphery, the culture, the people not visited lie just beyond a buffer between what will be allowed and what must be kept out the reach of the visitors' gaze. The landscape idea originated as the outsider's perspective; it remains a controlling composition of landscapes in third world contexts where they are created for first world consumption. Representations of tourist destinations rely on certain culturally based stereotypes and received images (Selwyn, 1996; figure 1). The decisions to include or to exclude as well as how to depict what has been retained are not haphazard. Eliminating or obscuring is fundamental to the western process of landscape creation. Heterotopic destinations are their material realisation and the social practices which they inform. Heterotopias are not a mirror of the locals' experience or bonding.

Colonial, and today tourist, narratives have strategically functioned to produce geo-political myths (Barthes, 1957) about destinations. The emphasis, in the tropics, is on sunshine (as in the Tuamotu Islands where 'it is never grey', according to *Rewa Magazine*, Air Tahiti Nui's publication), on a particular set of illusions that conflate weather and social practices, transforming tropical destinations into consumable paradises. It can only mean that other geographies (and their darker residents) have been displaced. Any remnants of non-western culture must disappear from the gaze of the vacationing visitor. They can only be reintroduced as exotic curiosities, as staged folklore... "in every case quaintly meek" (Urbain, 1996, p. 172, author's translation), reconstructed spectacles, a form of programmed exhibition of these simulated, 'exotic' cultures. Paradise lost must be recreated: the model is to evoke authentically traditional 'purity'.

Figure 1: Tubuai advertises its natural wonders and "traditional" heritage to potential visitors. There is no mention of the present day social and cultural production, source: author

Despite their standardised architecture and controlled environments, these heteretopias do try to showcase the local 'exotic' mystery and romance through visitor participation in staged performances, "a collective dramaturgical game in which guests fantasise and act out situations that they imagine but will never confront" (Edensor, 2001, p. 154). This imaginary is defined by semiotic codes and cultural themes of capitalist consumption. The performances are fuelled by colonial fantasies of exploration, of a desire for the sexualised other, of a yearning for the transgressive. Such staging serves to transform formally different cultural practices into seemingly familiar and intimate activities, a process of deworldment as it deprives local inhabitants of any rights. It is another illustration of where metaphor and analogy have led to inappropriate comparisons or rapprochements between incompossible categories and subjects.

Constructing exoticism

Tourism landscapes of the third world are constructed for their western consumers, who reside in them serially even as the tourist caravan only visits its preferred vacation spots (Zurick, 1995). Augé adds that tourism landscapes are a spectacle offering "a series of instantaneous visions which will not regain any reality until they are displayed again upon the tourists' return" (1997, p. 32, author's translation). Their natural beauty or some hybrid tradition is used as the attraction: many tropical resorts in Tahiti and its Islands have been built in coconut plantations, a form of cultivation imposed by colonial administrators (figure 2).

Tourism compresses and displaces down time and dead space to privilege immediate experience and adventure, rendering tourism landscapes as 'non-places',

Figure 2: Coconut trees, "traditional" materials and turquoise lagoons, source: author

i.e. spaces that prevent those who travel through from reading either their identity or their relation with 'others' (Augé, 1995) that Goss has so aptly described. They are but "the threshold between the past and the present, nature and civilisation, east and west, and sea and land" (Goss, 1993, p. 680), the result of unworldment processes, of the "undoing of landscape geographies"(Terkenli, 2002, p. 231). In other words, they really are nowhere, not even in a physical periphery.

Such margins are in fact defined by a powerful normative consensus that is located in the first world. They evoke serene tropicality to be shared by small intimate groups: a totally artificial world enclosed in its own bubble, an enclave. Goss confirms that "liminality ... is not about the experience of an ... Other, but an emancipation of the Self... it promises an encounter with the Self"(1993, pp. 681-2). These are apprehended as spaces empty of tension, places of (re)creation where one may recover one's potential as a sensitive individual, which was lost in the modern urban bustle (Barrell, 1980; Meethan 2001). The founder of Club Med was well aware of such a trope: "Gilbert Trigano noticed that most vacationers exhibited the desire to escape anguish and to desperately seek emotional and physical safety" (as cited in Peyre and Raynouard, 1971, p. 5).

The explorations that metropolitans seek in distant lands concern only personal metaphysical quests and displacements rather than sympathetic understandings of the spaces visited. Tourists imaginatively project their desires to constitute the landscape: "Tourism lies in the maximisation of pleasure, not achieved by 'useless wandering' ... [but] by temporarily inhabiting a destination explicitly designed for such a purpose" (Stewart & Nicholls, 2002). Tourism spatiality, which involves the western imposition of name and narrative, expresses the articulation between regimes of power, in favour of western(ers') pursuits (figure 3). There is no participating in the lived realities of the local people. The important link then is the

Figure 3: The local instructor awaits consultation between visitors and their willingness to buy his product. Sunshine, "traditional" materials and coconut trees frame these power relations, source: author

relations of power which continue to dominate the new cultural economy, in favour of the first world and its residents.

Travel today reaches ever further out, relentlessly encompassing new destinations in remote "peripheries," unexplored (by western tourists) and thus still shrouded in mystery. In Tahiti and its Islands they are now created in atolls that were still unknown to most tour organisers last year because ever further removed from the centre: Tikehau and Fakarava, for example, in the Tuamotu Islands of French Polynesia are the latest to offer four star lodging over clear lagoon waters, displacing Rangiroa and Bora Bora as remote hideaway paradises. The natural landscape and the materials used to construct the resorts confirm the liminality of the destination but the culture and the presence of the 'other' are markedly absent.

TROPICAL LANDSCAPES AS A SOCIAL AND ECONOMIC PRODUCT

The cultural economy is made up of 'relations of production' and 'conditions of reception', while it provides an "institutional framework that mediates between production and reception, and the particular way in which cultural objects circulate" (Lash, 1990). Culture is a dynamic process, a continual struggle over social relations, perpetually contested. It thus should never (have) be(en) separated from

politico-economic practices. Bartolovitch emphasizes that "the contest of cultures …
simply cannot be divorced from rigorous critique of the imbalances of global
political economy" (2002, p. 12). Globalisation could mean that "local
idiosyncrasies can thereby draw on a global resource as opposed to having a global
imperative subsume local idiosyncrasies" (Sluyter, 2002, p. 212). Globalisation,
however, is attributed a "cultural personality, a fixed disposition, a programmatic
way" (Lazarus, 2002, p. 53) that all necessarily but normally submit to while it
erases its socio-material basis. It is rarely described as the expansion of capitalism.

The new cultural economy asks us to contextualise materiality. Even postcolonial
theory had displaced or bracketed the specific agency of capitalist social relations in
imperialist development (Lazarus, 2002; 1999). Post colonialism, contends Dirlik
(1998, p. x), "relegates to an ideological or cultural Euro centrism the responsibility
for past inequalities and oppressions … that nourish off not just memories of past
inequalities, but their contemporary legacies". It is important to understand that
capitalism (not Euro centrism or 'cultural imperialism') underwrites those relations
of contemporary "uneven and combined development" (Bartolovitch, 2002).
Tourism has been incorporated into the political rhetoric of governments about
ending the evils of poverty but we need to remember that developmental (i.e.
productivist) ideology was part of the colonial state and its capitalist expansionism.
The result has been the creation of an acutely uneven development skewed in favour
of certain regions while it has increased the institutional and political marginalisation
of the poor.

One of the hidden purposes of the new global economy is to reduce spaces of
production by those who are not supposed to accumulate capital: In Wallis and
Futuna, westerners would describe the indigenous people as irresponsibly wasteful
when they displayed their wealth for special occasions like marriages, funerals or
baptisms: too many pigs slaughtered whose meat turned foul in the heat of the day as
ceremonies proceed ever so slowly! Yet waste for the purpose of greater
accumulation is one of the basic traits of capitalism. These exoticised fantasies of the
Third World obscure the continuous reality of the one integrated world of relations
between exploiter and exploited. Bauman underlines that "the market … hides the
practical inequality of consumers … the sharply differentiated degrees of practical
freedom of choice" (1990, p. 211). We need, however, to dispel the notion of a
linear, static and unequal set of power relations between one subaltern group and
another in political dominance (Loomba, 1998) since no one single process is at
work at any time.

Cosgrove asserts that "landscape represents a way in which certain classes of
people have signified themselves and their world … and through which they have
underlined and communicated their own social role and that of others" (1984, p. 15).
Some social groups have more power than others to determine what is desirable and
where. The use of leisure space involves forms of symbolic closure that precludes
some groups from access. Adapting the feminist argument that history was literally
'his-story', a narrative of masculinist domination, the negation of the beliefs and
actions of 'others' perpetuates the assertion of colonial ideology in today's capitalist
relations including tourism ones. Tourism has also continuously affirmed distinct
class divisions: its consumers, for example, recount specific travel narratives to

enhance *their* social (&/or cultural) capital. Post colonialism connects remarkably little with conventional developmental agendas (Simon, 1998), yet tourism is a type of economic activity that participates in the articulation and perpetuation of uneven development and of unequal territorial relations.

Tourism, as a circuit of capital, is to encourage the backward and isolated communities it circulates through to modernise, even though their attractivity resides in their supposed inability to progress. Many tourist sites have also been developed in areas still colonised. Such third world destination landscapes are thus made to represent a traditional rather than a modern economy, for the consumption of Western tourists anxious to take a break from 'modernity' even though "the tradition that is invoked to restrict activity … to the kind of activity which was common in the past is a false tradition" (Sharp, 1946). Western tourism discourse controls the production and dissemination of tourism, hence its nature and shape, action reinforced through tourist processes. In the Third World, tourist landscapes are thus most often moulded by other than local forces. They have been created for particular types of use by those who gain from such constructions and thus represent what foreign investors choose to present to future visitors.

These monuments to tourism are "deliberate physical manifestations of ideology, a more or less massive inscription of triumphal or laudatory statements upon the landscape" (Shurmer-Smith and Hannam, 1994, p. 203), in particular a form of triumphant capitalism "to a world no deeper or more important than that which people can construct" (Nozick, 1975). Gregory adds that these are "ideological landscapes whose representations of space are entangled with relations of power" (1995, p. 474). Foucault confirms that "power produces; it produces reality; it produces domains of objects and rituals of truth" (1979, p. 194). These sites look ideal and splendid as if there were no modern indigenous inhabitants but only Western tourist spectators whose corporeal presence remains the only way to experience these exoticised destinations despite ever growing numbers of representations and forms of depiction.

MATERIALITY OF TROPICAL DESTINATIONS

These self-contained enclavic destinations are disconnected from their surroundings and lie at the end of some 'extraordinary' journey. These phantasmagorical places are very concrete, not merely reflective imaginaries. They are built universes or places even if the materials are sometimes merely wood or plant rather than solid concrete (see Mitchell, 2001). Tropical landscapes are inscribed with these physical marks which are foci of western expression and thus mould and facilitate the transmission of western values and ideologies. They are not just images or representations of fantasies. Such liminal places are tempering zones lying between familiarity and proximity and the danger of truly remote places. The aim is to impress the visitor with the sensation of living an exotic adventure, to create an illusion of participation in the local culture without risk of being creolised or contracting an infectious disease.

The audience addressed shapes the statements that can be uttered. 'Others' have often been identified in colonial discourses with savagery, careless corporeal management, beyond control. They did not perform proper hygiene; they practised sexual laxity and displayed great moral disorder (Melville, 1849; Kingsley, 1897). Segregation, on health grounds but also to avoid miscegenation, was a current practice in colonial territories. Protective screens are still needed to separate civilised visitors from contamination. Tourists' encounters are "in great measure contained within deliberately designed sites and attractions" (Katriel, 1995, p. 280) to shield them from the strange and unexpected while exposing them to the desired "authentic" differences.

They are conceived as spaces of possibility – places of excitement, difference, adventure. They illustrate some of the material practices that appropriate and transform real places. The spectacular requirements of the users of the resort and not the sensibilities of the 'others', the residents of the host area, forcefully shape such management and arrangement. Nature too is not just for viewing. Tourists are recolonising it. On a visit to Borneo, tourists depicted nature there as "both wild and

Figure 4a: Young indigenous women dance in grass skirts: illustrating their closeness to nature, source: author

Figure 4b: Indigenous women are not conceived as successful entrepreneurs wired into global networks: banking profits while keeping in touch with a cell phone. source: author

yet tamed, or at least tameable" … by (and for) them (Markwell, 2000, p. 94). But such myths had already been fixed as European stereotypes as far back as the 1600s. Nature is contained and restrained (McCannell, 1992) in the same way as the movements and thus the experience of the visitors for whom an encounter with static traditions is safer than a confrontation with a dynamic present (figures 4 & 5).

The tourist is not a free subject of thought or action, nor is the host. The structuralist approach suggests hosts are people who have tourism done to them by guests but interactions between the two groups tend to be complex, dynamic and dialectic. Tourists are often mere consumers of a "culturally constructed hypersublime, not authors of their own unmediated interrogation of that new topography" (Bell & Lyall, 2002, p. 20). Some destinations have even been recast, not just as spectacle, but to serve as a 'greening of consciousness' that assuages the guilt of visitors without requiring any action on their part.

To reassign relevance to the visited site, these non-places, and to the visit, tourists collect 'mementos' in the same way as others leave objects to "establish the location of a place" (Richardson, 2001, p. 260) through their reflexive power. These artefacts (including photographs) recall and reconstitute the experience of presence, "instantaneous visions displayed upon the tourist's return" (Augé 1997, op. cit.), to conjure absence and death inscribed in 'non-places' or in vacant space. The minimal agency of visitors leads them to collect with great urgency these reassuringly material, solid, proofs of their passage and of their selves. Capitalism needs us to believe we travel to fulfil deep desires. It behoves it not to play its hand (an accelerated requirement to commodify experiences) too obviously.

Hybrid resorts that marginalise local hosts can only be constituted through the interplay of structurally articulated elements of experience of visitors who have no other commitment to those sites, so the "recounting of happenings" (Malpas, 1999) is one imposed on the very features and contours of the local landscape. Such imposition obliterates the original layerings of local cultural narratives that constitute(d) them and their attractivity. At the same time, minimal encounters with local practices or with their simulacra often encourage negative stereotyping of the spaces not visited. The roots of these stereotypes can be traced to colonial discourses and taxonomies used to describe difference. Science has often been used to legitimise such processes of inferiorisation, subordination and inequality. The reiteration of such discourses constructs regimes of truth and leads to their normalisation and acceptance (Foucault) so that observations made by travellers (including earlier ones) maybe confused with causation.

A POSTCOLONIAL PERSPECTIVE.

The legacy of colonialism is evident in many of the structures and practices of contemporary leisure and tourism. Their colonial roots lie in the material appropriation and use of local properties by destination developers and managers. They remain colonial as a comment on social relations between the West and the locals. The exhibition of 'natives' ('exotic people') in world expos of the nineteenth century was an exhibition of the nation that presented them. It was proof of the

Figure 5: When older, indigenous women create cultural artefacts, "souvenirs" for tourists, dressed in "traditional" costumes (missionary dresses imposed by colonialism, in designs reserved for specific territories), source: author

greatness and of the colonising, taming and domesticating power of the nation exhibiting them. The colonial world fairs were far more important when tourism was not so widely practised. The construction of tourist destinations often follows the same model of 'exhibition', of subservience, of primitiveness and tradition for the Western public.

At the same time tourism developers benefit from cheap labour and land for their resorts in tropical regions. These third world sites also offer investors other economic incentives. These resorts are symbolic shorthand for the manifest invasion by affluent visitors, and an extension of colonialism (Kincaid, 1988). Post colonialism has perpetuated those myths, this *imaginaire,* now consumed as (p)lei(a)sure destinations and not just as fantasy or simple escape reserved for the elite. The continued celebration of difference and otherness by tourism should be seen as a process similar to the one that supported empire building. Destinations today, just like colonised people of the past, consume narratives and images of themselves that are imposed on them. The discourse (construction) of tourism in the west determines its conditions of possibility. The production of tourism destinations, reified as paradise and commodified as hideaways, has responded essentially to the interests of the metropolitan centre.

Getting close to the locals obviously means decadence. Since s/he belongs to the heights, the tourist refuses to enquire about the fate of those below (Amirou, 1995). All travels bear traces of the desire to conquer the world (Todorov, 1987). In the sixteenth century, the term 'voyage' still often referred to a military expedition: "to travel was to cast a murderous gaze upon the world. War was considered a fundamental type of rapport with the 'other'" (Doiron, 1995: 159, author's translation). Some writers have asserted that God invests travellers with the right to take possession of the whole earth (Maritain, 1941). Native populations remain in colonised spaces so that for a space to become a resort, its original occupants must be removed. Their absence (and hence their dislocation) becomes naturalised through marketing discourses. The statements that make up these discourses "refer back to an institutional milieu ... and to certain intrinsic positions" ensuring their density of meaning and their validity to their readers (Foucault, 1986).

Travel guides support this selective aesthetic appreciation as they focus the traveller's gaze on particular features. They have perfected their practices since the sixteenth century even though their emphasis has shifted through time (Stagl, 1990). According to Barthes (1957), they effectively depopulate the landscape of real flesh and blood people, putting in place only representations of ideal types, stereotyped categories in which the part stands for the whole. Advertising offers us tropes, the trope of time travel which invites us to step back in time when we visit these 'traditional' peripheries. The value of 'the edge' is increased through remoteness and isolation as distance enables one to travel back in time. The trope of emptiness allows us to 'discover' new worlds, together with the trope of 'uncontaminated' traditional authenticity as tourists search for new, 'different localities'.

Post modern tourists wish to unfold their own personal daily praxis of such sites, each desiring that 'unique' experience but they may be faced with the perpetual disappointment that their encounters will never live up to their ideals. The process of colonialisation and imperialism prevents us from discovering or returning to a pure experience untainted by modernity and Western influences. Imagined or mythical understandings of 'other' people and places impact on the relationships between different societies (Rose, 1995). Said (1978) attempted to show that Western attitudes to the 'Other' were racist in that they relied on stereotyping of people and cultures in order to support a feeling of Western superiority. Colonial entanglements have produced a new 'soft' cultural imperialism that defines cultural authenticity whose worth is determined by its commodified value (price).

The tourist destination is stereotypically imagined as a sort of exotic paradise, a place of luxury, leisure, sexual pleasure, seductive perfumes and mysterious magic for visitors' enjoyment. The fantasy of an erotically exotic female 'other' became a vehicle for Western libidinal fantasy (figure 4a). Warmth and nakedness had combined to create the illusion of sexual freedom and free sex, providing the perfect breeding place for white men's dreams (Woods, 1995; Faessel, 1995). The space or life-world they displace, however, is "not only the field where individuals' existence unfolds in practice [and could thus be removed at will to a different locale]; it is where they exercise existence" (Mbembe, 2001, p. 15). The production of everyday life and all its particulars is linked to spatiality: colonial processes of inscription continue to be manifest today in the third world. The presence of colonisation remains inscribed in third world tourism spaces through these material constructions and their representations.

Destinations need to offer opportunities for metaphysical displacement which also means that local experience no longer coincides with the place in which it takes place, particularly for third world hosts of first world guests. Tourist resorts are lived as outposts of Western civilisation from which to consume the 'other' culture through a romanticised interpretation or references to its ancient past (figure 1). Few featured activities are truly 'primitive'. Colonisation, through powerful means of conversion (including religious), has allowed few to survive. Today's traditions are colonial hybrids. Tourism subtracts much of the complexity of places, simplifying and decontextualising it (Kirschenblatt-Gimblett, 1998) for the easy consumption of tourists in quest of timeless and unchanging oases of difference, thanks to a 'miraculous evaporation of theory' (Barthes, 1957). Baudrillard has underlined that one never cognises the world as it is but rather recognises the world as its own proper image and correlate (1983). The location becomes almost irrelevant (Kitchenblatt-Gimblett, 1998).

TROPICAL TOURIST LANDSCAPES AND THE GAZE

Just gazing ...

Although Kirshenblatt-Gimblett (1998, p. 54) worries that "whole territories become extended ethnographic theme parks ... where the viewer sees without himself (sic!) being visible", the 'theme park' concept best expresses the static nature attributed to the 'exotic' landscapes sought by tourists. The story is more about the action of the tourist who is permitted to gaze upon without truly seeing. The panoramic notion of landscape has prevailed in the construction of these destinations. The touristic gaze, from the heights of its civilizing and technological pedestal, is a 'vertical' gaze, from higher to lower reaches (Amirou, 1995). Such a gaze distances the onlooker, transforming the landscape into an object that is separate and can thus be exploited for the gazer's ends.

In spite of the importance of 'ocular' metaphors and practices, or maybe because of it, and although the 'gaze' of tourists is governed by the dynamic of recognition, one of its main attributes is its blindness. It is blind "because it is unable to see beyond the law of visibility" (Beadsworth, 1996, p. 94) as in, for example, first world societies exporting of unseen violence to open an ever wider touristic periphery (Wheat, 2002; Mclaren, 1998; *Christian science monitor* video on Africa, 1993). Beardsworth (1996) argues that how we distinguish between the visible and invisible is first and foremost a socio-political question. "Visual experience is not self-identical or fully present" (Macphee, 2002, p. 11) since it relies on an invisible frame that organises how we see. Already in the nineteenth century, guide-books and instructional manuals tended to govern the inquiries of travellers and to define their field of vision (Barthes, 1957; Stagl, 1990).

Prejudices are involved in the every day act of seeing, where we are certain of what we see, while we forget the particularity of our own perspective and its part in shaping what is seen (see Doiron, 1995). Travel advertising is thus an agent of blindness. Photographs and tourist viewpoints have been 'invisibly' incorporated into capital circuits through commodification by, here again, obfuscating the means of such production. Adams (2001, p. 188) underlines that "ideologies ... have shaped this socio-spatial evolution" and that "the reduction of multisensory experience ... is fundamental to the loosening of ties" since it "reduces the surrounding landscape to scenery", to a panorama, an objectified 'reality' commodified for visitors. Their fetishistic desire, stimulated by advertising, disregards the original local Indigenous reality. It emphasises simulacra, a simulated reality which has no original (Baudrillard, 1983).

It encourages "a sterilized 'scopic regime' [that] abstracts and devalues both physical and social places, reducing the world to an image. In this visual space, the place-self relationship already loosened by the increasing *amount* of mobility [which opened long-haul destinations] is further weakened by the *nature* of the mobility" (emphasis in original; Adams, 2001, p. 191). The social capital of tourists amounts to successive, limited viewpoints, or visual appropriations of landscapes. What Alpers points out about photographs, which present characteristics that are "common also to the northern descriptive mode: fragmentariness; arbitrary frames; ...immediacy" (1984, p. 243), applies to postcards too. They are held as evidence of "a larger truth ... which might not be true at all" (Dyer, 2003).

... or appropriating?

Vision, as Macphee (2002) reminds us, was never divorced from technology so that modern conceptions of vision remain embedded in the evolution of global technology even as they direct its applications. The metaphor Kepler referred to by using the word 'pictura' to describe the inverted retinal image underlines how our view of the world is socially constructed. It also encouraged the modern understanding of vision as actively constituted by the observer. Technology has the capacity to reorganise the coordinates of visual experience (though it does not drive it: social and cultural choices do) so that we might question the forms of vision tourists and developers are seeking in these liminal destinations. In the nineteenth century, already, photography "was understood to be the agent par excellence for listing, knowing and possessing, as it were, the things of the world" (Solomon-Godeau, 1991, p. 155).

Comolli writes that "[through] a geographical extension of the field of the visible and the representable: by journeys, explorations, colonisations, the whole world becomes visible at the same time that it becomes appropriatable" (1980, p. 122-3). The view is organised in the best colonial tradition by treating the area as exhibit while the tourism industry invents ways to gaze upon these exhibits. Landscapes, originally conceived as paintings to be contemplated (panoramas), are now decoded as bodies or systems to be commodified. Photography maintains the circularity of the tourism system because the photograph provides the illusion of an objective rendering of reality. In these spaces of constructed visibility, how things are made visible materialises and gives legitimacy to the power and prominence of tourists (figure 3).

In the libidinised economic order (Lyotard, 1993) that exists today the form of exchange relations leads to the erasure of the production process and its corollaries. Privileging the appearance of commodities (including tourism landscapes) dehistoricises them and by extension relieves their consumption (consumers) of any guilt (Urbain, 1996). An invigorated sense of looking at tourism landscapes as dioramas thus remains even as all signs of the contemporary inhabitants have been erased (sometimes literally, Wheat, 2002) or rendered invisible, in particular in those landscapes that have been drawn into the abstract grid of neo-colonial power. The indigenous localities are scarcely ever named or given density, in the tradition of Locke's references to the "wild woods and uncultivated wastes of America", which formed, as O'Neill reminds us "part of the denial of rights in land to the Aboriginal population" (2002, p. 37).

RESISTING DISPLACEMENT

Tropical tourist destinations have become sites of appropriation and domination as power has displaced and replaced the original landscape with a consumer friendly exoticism made of lonely mementoes or fragments of the original, a 'performed' tableau vivant. These centres of human existence are hemmed in by large-scale social, political and economic structures. The aim is not to diagnose them as false or as 'bogus resorts' in the way Hewison has decried heritage as "bogus history" (1987, p. 144) but to highlight the vested interests involved in their promotion and the lack of attention paid to their social, political, environmental and economic impacts on hosting individuals and communities. Foucault thus comments that "…history serves to show how that which is has not always been; i.e. that the things which seem most evident to us are always formed in the confluence of encounters and chances, … and that since these things have been made they can be unmade, as long as we know how it was that they were made" (1988, p. 36-37).

Promoters insist that their intention is homage, not exploitation but, in fact, these landscapes represent sites "at which particular spatialities are captured, displaced and hollowed out" (Gregory, 1995, p. 477). Displaced people struggle to regain their own places while colonised people seek to make their own places rather than have them made for them. Foucault stressed that no spaces exist outside of power relations and power is dispersed through a variety of everyday spaces. Although we eventually normalise representations that are imposed on us, Harvey has observed that "struggles over representation are as fundamental to the activities of place construction as bricks and mortars" (1990). Representations do not simply reflect a state of things. They actually produce realities of dominance and subordination. Exploration and imaginative appropriation remain integral to the visit since these destinations can be characterised as discrete and controllable places of otherness.

The impossibility of 'dwelling' in these destinations for the local populations leaves little opportunity of rejecting or even transforming such physical inscriptions and the mode of domination they represent. Communities are not overlapping; there is no shared political arena. These destinations are a version, to some extent, of the 'scenic type of landform' described by Olwig (2002): their essential scenic unity obscures the diverse global as well as regional and local actions that have created them. It also erases the local social order that previously occupied that space. They are no longer a location in which local social life is actively engaged. Negotiation cannot happen since all parties need to feel they participate in the arrangement. Locals are not free to refuse to consent to the construction of such landscapes.

Boast (1997, p. 188), however, argues that these material inscriptions become "actor[s] in their own right, being delegated entities, roles and social status dependent on their constitution within the heterogeneous network". In such a dialectic process the material world cannot over determine the social world. Its presence interacts with local experience and local residents should in turn be able to imprint it. They should retain some agency, even in their subservient roles as relations are never static but in a constant state of becoming. Social practices beyond the limits of the resorts could be co-opted in this form of resistance: "resistance is the main word, the key word, in this dynamic" (Foucault, 1997, p. 167) if such could be acted upon.

The question remains about how to resist when goods and services developed and sold by the periphery have little success in penetrating the centre (Lang & Hines, 1993, p. 15). Lapham confirms this feeling of powerlessness as poorer countries "become colonies not of governments but of corporations ... and the world's parliaments intimidated by the force of capital in much the same way that ... they had been intimidated [earlier] by the force of arms" (2004, p. 167). Some authors would insist that rather than searching for community resistance, we should consider the daily practices of individual lives (Cosgrove, chapter).

From the Foucauldian perspective enclavic destinations (heterotopias), or fantasy-architectures, become places of constant surveillance where the elimination of 'distasteful' bodies (local people or tourists of a different class) is as important as eradicating abnormal acts (e.g. non-heterosexual forms of behaviour or non-western cultural practices). They remain a showcase for public entertainment and education designed to keep people, in particular natives, in their place, which requires the everyday management of bodies in the layout and arrangement of space. Colonised subjects have become objects to be displaced at the whim of visitors (and those who profit from their practices). These destinations occupy space that has been deterritorialised. They have lost their quality of 'place' as traditional social bonds are compromised by foreign ownership based on profit. The new cultural economy has rescaled this economic space through the connections and interactions that have been established to ensure profitable returns for its investments.

NON-HETEROTOPIC SPACES

A few destinations are being allowed to develop as 'post traditional' or non-heterotopic destinations. It is hypothesised that these tourism hybrid landscapes are actively constructed by local actors but also still by tourists of the first world. The experience of Nepal has shown that tourism and conservation can reinforce each other especially if strategies and processes to ensure gains to local communities are put in place (ESP, 2004). The spatial explorations that some metropolitans embark upon concern understandings of the spaces visited, a desire to participate in the lived realities of the local people. Alternative tourists, according to Game, would "practice a way of being-in that inscribes a different desire" (1991). They would be in search of landscapes or destinations that integrate locals and are inclusive of their life's occurrences for their 'authentic' difference.

Indigenous destinations offer opportunities for metaphysical displacement as their indigenous developers both rediscover their past and invent their future. It is from and within history that places are imbued with value. The real and not real, the actual and the fictional are not necessary distinctions in the same way as backstage and front stage are not necessarily distinguishable in third world countries. They are imaginatively bound, signifying potentially the quality of a revelation for the visitor. Place is a texture of relations (Rehmann-Sutter, 1998), including historical and traditional ones. Such 'lived experience' can prove transformative for visitors who recognise how their knowledges of other cultures are constructed but also fractured. We need to be careful, though, not to romanticise notions of authenticity or tradition.

Indigenous owner-operators should be able, within their own spaces, to more directly demonstrate their culture to visitors even if normality, for their visitors, is still defined by the comfortable presence of fellow white tourists with whom they share experiences and swap itineraries (Edensor, 1998, p. 28). Hybridity here does not signal an uncritical celebration of cultural syncretism or an aestheticisation of politics. We need to recognise the ability of third world people to consciously choose to act as agents. It is also a way to rescue colonial subjects from perpetual victim hood. Problems of appropriation still exist in these instances of cross-cultural exchanges but they also make possible new forms of communal or individual practices as the introduction of different technologies enable people to act in novel ways.

Fetishism, however, often afflicts such destinations as they insert themselves in the global circuits of advertising. Some visitors want to believe that encounters with Natives are spiritually defined and balk at the local need for moneyed transactions. Such commercial concerns are believed to diminish the 'authenticity' of the other (Edensor, 1998). The experiential diversity of encounters with difference does provide visitors with cultural capital that will be moneyed upon their return to the West. These encounters thus perpetuate a colonial construction of non-Western spaces to be used as a stage for western fantasies and desires. In both heterotopic and non heterotopic destinations, control is effectively exercised through Western colonialist attitudes towards 'others'.

These non heterotopic (run by indigenous or native people) destinations face some of the same issues as 'nature' ecotourism destinations (often but not always run by foreign entities). Attributing the notion of 'authentic tradition' to tourism destinations erases the connections and interrelations they maintain with the global: religious communities, supply chains (import-export ones, for example), cultural ties, other regional or global links and migratory trajectories. Such a notion erases their varied network typologies of dynamic agents. It also constrains their identity which can only be anchored in a frozen, static past.

They also exist within a global capitalist context. All places evolve within global and regional economic and cultural relational networks which shape their spatiality and which can cast doubt on their ability to provide significant local returns. Non-heterotopias must be situated, paying attention to the economic, political and social inequalities of the environment within which they must be developed and nurtured. In a relationally constituted global, cultural economy, even when run by third world agents, no area can be circumscribed for effective local economic control. No local initiatives can maintain locally bounded growth and returns without a reduction or reordering of international economic power.

Non-heterotopic destinations, just like heterotopias, exist to attract visitors and their disposable income: their primary purpose is not to educate, whoever might be constructing them, even if tourists accept the guilt-ridding educational discourse that permeates the marketing of such forms of tourism. These destinations are all in the business of selling culture, pleasure, fantasy, and spectacle though in which order is difficult to untangle. Defining what 'culture' is for sale has suffered from paternalistic attitudes from national governments or international bodies even when engaged in creating such 'sustainable indigenous spaces'. Even non-heterotopias find themselves selling a 'traditional' rather than a 'modern' representation of themselves. They have to commodify experiences culturally ordered by the west to engage with the new global cultural economy. They have to dramatise their potential for investment before they can perform as profitable enterprises. It remains a form of colonial linear economic development imposed on these 'remote' peripheries.

CONCLUSION

Tourism and tourist practices are 'a manner of being' (Foucault, 1986, p. 377). Subjectivity, according to De Laurentis, is determined by experience, "the continuous engagement of a self or subject in social reality" however that social reality is constructed (1984, p. 182). Subjectivity gives individuals agency even if it is constrained within particular discursive configurations or domains of visibility. Tourists are thus active participants in the creation of these tourism destinations as they maintain a monologue from outside. They are insiders who have displaced the locals to outsiders within their own locales. As in colonial time, there is a "forced, often violent production of an abstract(ed) space" superimposed over the particularities of the place it represents but has rendered invisible. These hybrid landscapes express the interconnection of vision and appropriation by (neo)colonialism and the continued 'displacement' effected by modernity (Giddens, 1990).

An active participant network creates (designs and develops) and consumes tourist destinations in the third world. These destinations are 'quasi-objects' that originate in the Western (first) world and transform the landscapes of the third world. These quasi-objects are grafted onto the local landscape with little regard for their impacts. The 'hosts' of the tourists are Western developers, investors and managers. Local people are recruited to participate as peripheral elements, rather than agents, in the entertainment of the 'guests'. The 'quaintness' of such landscapes or entertainment only serves to emphasize the areas' present political and economic subordination to processes of unworldment and deworldment. They exist only as "a surface on which to set forth or inscribe the world rather than as a stage for significant human action" (Alpers, 1984, p. 137). In these places of constructed visibility, such agency materialises and gives legitimacy to the power and prominence of tourists.

Tourists behave as 'insiders' within the places (destinations) they have created, which has meant displacing the local landscape, culture and its owners into 'non-places'. Such enclaves "contextualise and tame the difference, emplacing it in a predictable epistemological setting" of tourist narratives (Adams, 2001, p. 201). It is yet another form of colonial violence that our forms of seeing enable us to ignore as we enjoy the 'elaborate exoticism' of our chosen destination. To resurrect these landscapes from their status as 'non-places', tourists bring back mementos that recall and reconstitute experience in opposition to absence and death inscribed in 'non-places' of unworldment.

Such constructed spaces remain fragile, vulnerable (see Malpas, 1999). Place is described as dynamic and fluid, as a 'contested terrain'. "The dominant image of any place will be a matter of contestation and will change over time" asserts Massey (1994, p. 121). Recovery of an effective identifying relationship between self and place for the original occupants and their construction of cosmopolitan destinations are yet possible. Social relations are unendingly heterogeneous, not fixed, possibly moving away from deworldment to transworldment. We need to search for new foundations upon which to reconstruct 'authentic' landscapes. These new contexts would facilitate the production of social meanings and practices to constitute places that can incorporate difference to be explored and appreciated by tourists.

ACKNOWLEDGMENTS

The author wishes to acknowledge grants received in 2001 and 2003 from the University of Waikato as well as a travel grant from the government of French Polynesia in 2001 for research in the French Pacific, which much of the work of this chapter is based on.

REFERENCES

Adams, P. (2001). Peripatetic imagery and peripatetic sense of place. In Paul Adams, Steven Hoelscher and Karen Till (Eds.) *Textures of place: Exploring humanist geographies* (pp. 186-206). Minneapolis: University of Minnesota Press.

Alpers, S. (1984). *The art of describing : Dutch art in the seventeenth century.* Chicago, University of Chicago Press.

Amirou, R. (1995). *Imaginaire touristique et sociabilités du voyage.* Paris: Presses Universitaires.

Augé, M. (1995). *Non-Places : Introduction to an anthropology of supermodernity.* Translated by John Howe. London: Verso (c.1992).

Augé, M. (1997). *L'impossible voyage: le tourisme et ses images.* Paris: Editions Payot.

Barrell, J. (1980). *The dark side of the landscape.* Cambridge: Cambridge University Press.

Barthes, R. (1957). *Mythologies.* Paris : Seuil.

Bartolovitch, C. (2002). Introduction: Marxism, modernity and postcolonial studies. In Crystal Bartolovitch and Neil Lazarus (Eds.), *Marxism, modernity and postcolonial studies* (pp. 1-17). Cambridge: Cambridge University Press.

Baudrillard, J. (1985). *America.* Trans. Chris Turner. London: Verso.

Baudrillard, J. (1983). *Simulations.* New York: Semiotest(e).

Bauman, Z. (1990). *Thinking sociologically.* Cambridge, MA: Blackwell.

Beardsworth, R. (1996). *Derrida and the political.* London: Routledge.

Bell, C. and Lyall, J. (2002). *The accelerated sublime.* Westport, Ct: Praeger.

Boast, R. (1997). A small company of actors: a critique of style. *Journal of Material Culture,* 2 (2), pp. 173-198.

Clark, H. (1993). Sites of resistance: place, "race" and gender as sources of empowerment. In P. Jackson and J. Penrose (Eds.), *Constructions of Race, Place and Nation* (pp. 121-142). London: UCL Press.

Comolli, J-L. (1980). Machines of the Visible. In T. de Laurentis and S. Heath (Eds.), *The cinematic apparatus*. London: MacMillan.

Cosgrove, D. (1984). *Social formation and symbolic landscape*. Madison: University of Wisconsin Press.

Crang, P. (1997). Introduction: cultural turns and the (re)constitution of economic geography. In Roger Lee & Jane Willis (Eds.), *Geographies of economies* (pp. 3-15). London: Arnold.

Crary, J. (2000). *Techniques of the observer: On vision and modernity in the nineteenth century*. Cambridge: MIT Press.

Cresswell, T. (1996). *In place/out of place*. Minneapolis: University of Minnesota Press.

De Laurentis .(1984). *Alice doesn't.*. Bloomington: Indiana University Press.

Doiron, N. (1995). *L'art de voyager*. Montréal : Les Presses de l'Université Laval.

Dirlik, A. (1998). *The postcolonial aura*. Boulder, CO: Westview Press.

Dyer, G. (2003, Jan30-Feb05). Dispatches from the edge of death, *Guardian Weekly*, p. 17.

Ecotourism Society of Pakistan (ESP). 2004. atlas-euro.org – list. News report South Asia losing share in world tourism, 28-09-2004.

Edensor, T. (1998). *Tourists at the Taj*. London: Routledge.

Faessel, S. (1995). La femme dans l'idylle polynésienne. In S. Faessel (Ed.) *La femme entre tradition et modernité dans le Pacifique Sud*. Paris : l'Harmattan.

Foucault, M. (1997). Sex, power and the politics of identity. In P. Rabinow (Ed.), *Michel Foucault: Ethics – subjectivity and truth* (pp. 163-173). New York: The New Press.

Foucault, M. (1988). Critical theory/intellectual history. In L. Kritzman. (Ed.), *Michel Foucault: politics, philosophy, culture* (pp. 17-46). London: Routledge.

Foucault, M. (1986). Politics and ethics: an interview. In P. Rabinow (Ed.), *The Foucault reader* (pp. 373-390). Hammondsworth: Penguin Books.

Foucault, M. (1979). *Discipline and punish: the birth of the prison*. Trans A. Sheridan. Harmondsworth: Penguin Books.

Game, A. (1991). *Undoing the social: towards a deconstructive sociology*. Milton Keynes: Open University Press.

Giddens, A. (1990). *The consequences of modernity*. Cambridge: Polity.

Gregory, D. (1995). Imaginative geographies. *Progress in Human Geography* 19 (4), pp. 447-485.

Goss, J. (1993). Placing the market and marketing the place: tourist advertising of the Hawaiian islands. *Environment and Planning D: Society and Space*, 11, pp. 663-688.

Harvey, D. (1989). *The condition of postmodernity*. Oxford: Blackwell.

Hewison R. (1987). *The heritage industry*. Andover: Methuen.

Hobsbawn, E. (1998). The nation and globalisation. *Constellations* 5 (1), pp. 1-9.

Katriel, T. (1995). From « context » to « contexts » in intercultural communication research. In R.L. Wisemand (Ed.), *Intercultural communication theory*. Thousand Oaks, CA: Sage.

Kincaid, J. (1988). *A small place*. London: Virago Press.

Kingsley M. (1897). *Travels in West Africa: Congo Français, Corisco and Cameroons*. London: Macmillan.

Kirshenblatt-Gimblett, B. (1998). *Destination culture. tourism, museums and heritage*. Berkeley: University of California Press.

Lang, T. and Hines, C. (1993). *The new protectionism: protecting the future against free trade*. New York: New Press.

Lapham, L. (2004). The agony of Mammon. In A. Amin & N. Thrift (Eds.), *Cultural economy reader*. London: Blackwell.

Lash, S. (1990). *Postmodernist sociology*. London: Routledge.

Latour, B. (1993). *We have never been modern*. Cambridge: Harvard Univeristy Press.

Lazarus,. (2002). The fetish of the West in postcolonial theory. In Crystal Bartolovitch and Neil Lazarus (Eds.), *Marxism, modernity and postcolonial studies* (pp. 43-64). Cambridge: Cambridge University Press.

Lazarus, N. (1999). *Nationalism and cultural practice*. Cambridge: Cambridge University Press.

Lefebvre, H. (1991). *The production of space*. Translated by D. Nicholson-Smith. Oxford: Blackwell.

Loomba, A. (1998). *Colonialism/postcolonialism..* London: Routledge.

Lyotard, J. F. (1993). *Libidinal economy.* Trans. Iain Hamilton Grant. London: Athlone.

Macphee, G. (2002). *The architecture of the visible.* London: Continuum.

Malpas, J, E. (1999). *Place and experience: a philosophical topography.* Cambridge: Cambridge University Press.

Maritain, J. (1941). *Le crépuscule de la civilisation.* Montréal : Edition de l'Arbre.

Markwell, K. (2000). Photo-documentation and analyses as research strategies in human geography. *Australian Geographical Studies,* 38 (1), pp. 91-98.

Massey, D. (1994). *Space, place and gender.* Minneapolis: University of Minnesota Press.

Mbembe, A. (2001). *On the post colony.* Berkeley: University of California Press.

McCannell, D. (1992). *Empty meeting grounds: the tourist papers.* London: Routledge.

McLaren, Deborah.(1998). *Rethinking tourism and ecotravel.* West Hartford, Ct: Kumarian Press.

Meethan, Kevin. (2001). *Tourism in global society: place, culture, consumption.* Basingstoke: Palgrave.

Melville, E. H. (1849). *A residence at Sierra Leone,* edited Hon. Mrs Norton. London: John Murray.

Mitchell, D. (2001). The lure of the local: landscape studies at the end of a troubled century. *Progress in Human Geography,* 25 (2), pp. 789-796.

Nozick, R. (1975). *Anarchy, state, Utopia.* Oxford : Blackwell

Olwig, K. (2002). *Landscape, nature, and the body politic: from Britain's renaissance to America's new world.* Madison: University of Wisconsin Press.

O'Neill, J. (2002). Wilderness, cultivation and appropriation. *Philosophy and Geography,* 5 (1), pp. 35-49.

Peyre, C. et Y. R. (1971). *Histoire et légendes du Club Méditerranée.* Paris: Le Seuil.

Rehmann-Sutter, Ch. (1998). An Introduction to places. *Worldviews: Environment, Culture, Religion,* 2, p. 175

Richardson, M. (2001). The gift of presence. In Paul Adams, Steven Hoelscher and Karen Till (Eds.), *Textures of place: exploring humanist geographies* (pp. 257-272). Minneapolis: University of Minnesota Press.

Rose, G. (1995). Place and identity: A sense of place, in D. Massey and P. Jess (Eds.), *A place in the world? Places, cultures and globalisation.* Milton Keynes: Open University Press.

Said, E. (1978). *Orientalism.* New York: Vintage Books.

Schama, S. (1995). *Landscape and memory.* New York: Alfred A. Knopf.

Selwyn, T, ed. (1996). *The tourist image: myths and myth-making,* New York: John Wiley and Sons.

Sharp, T. (1946). *The anatomy of the village.* Harmondsworth: Penguin Books

Shurmer-Smith, P. and Hannam, K. (1994). *Worlds of desire, realms of power: a cultural geography.* London: Arnold.

Simon, D. (1998). Rethinking post-modernism, post colonialism and post traditionalism: South-North perspectives. *Environment and Planning D: Society and Space,* 15 (1), pp. 55-72.

Sluyter, A. (2002). *Colonialism and landscape.* New York: Rowman and Littlefield.

Solomon-Godeau, A. (1991). A photographer in Jerusalem, 1855: Auguste Salzmann and his times. In *Photography at the Dock: essays on photographic history, institutions and practices* (pp. 150-168). Minneapolis: University of Minnesota Press.

Stagl, J. (1990). The methodising of travel in the sixteenth century. *History and Anthropology,* 4, pp. 303-338.

Stewart, R. S. and Nicholls, R. (2002). Virtual worlds, travel and the picturesque garden. *Philosophy and Geography,* 4 (1), pp. 83-89.

Terkenli, T. (2002). Landscapes of tourism: towards a global cultural economy of space? *Tourism Geographies,* 4 (3), pp. 227-254.

Todorov, T. (1982). *The conquest of America.* Trans. Richard Howard. New York: Harper & Row.

Urbain, J-D. 1996. *Sur la plage.* Paris: Payot

Wheat , S. (2002, June 20-26). Visiting disaster. *Guardian Weekly,* p. 28.

Williams, R. (1977). Plaisantes perspectives. Inventions de paysage et abolition du paysan. *Actes de la recherche en sciences sociales.* Paris: MSH-EHESS.

Woods, G. (1995). Fantasy islands: popular topographies of marooned masculinity. In David Bell and Gill Valentine (Eds.), *Mapping desire: geographies of sexualities* (pp.126-148). London: Routledge.

Zukin, S. (1995). *The cultures of cities.* New York: Blackwell.

Zurick, D. (1995). *Errant journey.* Austin: University of Texas Press.

KENNETH R. OLWIG

GLOBAL GROUND ZERO: PLACE, LANDSCAPE AND NOTHINGNESS

"I am moved by the work of the sculptor Robert Smithson: the map he makes out of a heap of broken clear glass, or his vanishing points that do not vanish, or his mirrors "displaced" in the landscape. He once photographed rocks in situ, then removed the rocks and photographed the holes in the ground – absent presence."

Annie Proux, *Big Skies, Empty Places* (Proux, 2001)

Milieu Festival, Rådhuspladsen, Copenhagen, May 2005, source K.Olwig.

T.S. Terkenli & A-M. d'Hauteserre (eds.), Landscapes of a New Cultural Economy of Space, 171-192.

Abstract:

Michel de Certeau's work on the concepts of *space* and *place* – though often mistranslated – provides a provocative means of linking these two concepts to that of the *landscape* of places. De Certeau's landscape analysis takes its point of departure in the contrast between the maplike global view of Manhattan's locations from the World Trade Center and the place generating practices of the pedestrians in the landscape below. The WTC's immediate predecessor as Manhattan's most monumental elevation, the Empire State Building, celebrated New York State's role as the entryway to America's western Empire. The World Trade Center was as global in its imperial aspirations as the Empire State Building was continental. The reduction of the WTC to "Ground Zero," however, gave the vanished structure a world of meaning through the empty hole of its structural antithesis, a place marked by nothingness. This paper will confront the theoretical perspectives presented to de Certeau's analysis of space, place and landscape seen in the light of the transformation of the WTC's monumental spatial erection into a cavernous zero.

GROUND ZERO

The French historian and anthropologist Michel de Certeau introduces his famous essay "Walking in the City" with the words: "Seeing Manhattan from the 110th floor of the World Trade Center" (1984, p. 91). To de Certeau the WTC was a quintessential "location" – "*lieu*" in the original French – in a geometricized urban space. de Certeau is specifically using "lieu" analogously with what the philosopher Merleau-Ponty termed "'geometrical' space ('a homogenous and isotropic spatiality')" (Certeau (de), 1984, p. 97). The WTC, according to de Certeau, is "the most monumental figure of Western urban development" (1984, p. 93). It is remarkable, almost prophetic, that de Certeau begins on the 110th floor of the WTC because his essay is about the "transgressive" "emptying out" of such locations in space, by which they become transformed into places in a "landscape" ("*paysage*"). The subsequent violent transgression of the WTC by terrorists reduced this monumental location to a place of enormous significance to the political landscape of not only New York, but the globe by making it a "ground zero," or, as the dictionary defines it: "1 : the center or origin of rapid, intense, or violent activity or change 2 : the very beginning : SQUARE ONE *herein*" (Merriam-Webster, 2000: ground zero). To call such a hole "zero" is especially apt if, "zero's hollow circle had its origin" in the "shallow depressions" originally left by counters placed in the sand by Arab or Indian merchants, as Robert Kaplan hypothesizes in the historical study, *The Nothing that Is*. These then became "the places" that "counters lay in or move through" in counting boards and board games," so that "an empty pit would be the aptest symbol for zero" (Kaplan, 2000,

p. 92). This essay is about the easily overlooked significance of an absence that is present, or of a nothingness that is something, in short, the importance of zero to the understanding of place, landscape and the world.

A "Quantitative" Theory of Place and Landscape

Place is difficult to define, perhaps because its widespread popularity in fields such as geography, architecture and planning came about largely as a critique. Functionalistic spatial planning and quantitative locational analysis were seen, according to the critics, to create "placelessness" (Relph, 1976). Place was thus a positive that was defined in counter distinction to the negativity of functionalist space. Place was the antithesis of location in space, which was the focus of the quantitative geographers. As Yi-Fu Tuan put it:

> As location place is one unit among other units to which it is linked by a circulation net; the analysis of location is subsumed under the geographer's concept and analysis of space. Place, however, has more substance than the word location suggests: it is a unique entity, a 'special ensemble' . . . ; it has a history and meaning (Tuan, 1974, p. 213).

Location and *place*, however, often appear to be used synonymously, and this raises the question of exactly how they differ. Is place simply a location, or portion of space, that has been incorporated into "the local 'structure of feeling'," as suggested by the British dictionary of geography (Johnston, 2000: place, 358)? Is *place*, thereby, the subjective, ideographic counterpart to an objective, nomothetic, quantifiable location?

The alternative to quantitative/qualitative, objective/subjective binaries is, I would argue, to examine the concept of place on the terrain of the quantitative and mathematical itself. *Zero*, I will argue, has a similar enigmatic quality in mathematics as that of *place* in geographical discourse, a situation that may be better understood when examined in light of the historically important role that mathematics has played in geographical discourse going back at least until Ptolemy. If place is conceptualized as zero, I will argue, it cannot simply be reduced to the qualitative counterpart of location – a point on the correlates of the map. A point in the mathematical geometry of the map's grid is, in principle, so infinitely small that it belongs to the realm of the ideal (Edgerton, 1975, p. 81). Zero on the other hand, as Robert Kaplan explains, is a "nothing that is" (Kaplan, 2000). It is, as zero's biographer, Charles Seife, explains, a "placeholder" for something (Seife, 2000, pp. 15-16). The concepts of zero and place are inextricably linked in the notion of zero as a placeholder, place thus becoming definable as that which is held by a zero. A place, unlike a point on a grid, cannot be infinitely small, because there must be room for something to take place. The concept of zero is inherently enigmatic because it is notoriously difficult to reason about a nothing that is – an absent presence – and this has been true from the earliest times (Kaplan, 2000; Seife, 2000). All this talk of nothings that are – and presences that are absent –, seems, on the surface, to be very esoteric, and for this reason, I will begin, via de Certeau, by examining, a concrete and familiar phenomenon from everyday life that goes under the name of place, and which appears to have given its name to the concept of place.

PLACE DE LA CONCORDE AND RÅDHUSPLADSEN

According to de Certeau, places such as the "*Place de la Concorde*" are created in a process by which locations, such as a cross roads, are transformed through forms of transgressive practice by which they are "emptied out," thereby "making room" for a "void" or "nowhere" that is liberated as "an area of free play" so that it can be filled with passages of meaning; an emptiness that can be "made habitable" (Certeau (de), 1984, pp. 104-106). The concepts of emptying out, transgression and zero, in relation to *place*, are, however, rather abstract and imponderable. It is thus better to begin an essay on this subject by starting with the concrete. Actual places, such as *Place de la Concorde*, are perhaps a particularly appropriate place to start because, etymologically speaking at least, such places provide the origin of the word *place*: "Middle English, from Middle French, open space in a city, space, locality, from Greek *plateia* broad street" (Merriam-Webster, 2000, place). It should be noted that the word "plaza" has the same etymological origin, and is applied to the same sort of phenomena as the "*place*" in *Place de la Concorde*, or in "market place" (Merriam-Webster, 2000, plaza). In cities laid out according to a grid plan such a place has often been squared, in which case a place may be known as a "square."

A French person, coming to the rigidly gridded and striated space of modern New York City, would naturally be struck by the contrast to the fluid topology of the cores of ancient European cities, where public life centers on one or more open "places." New York, to be sure, has places of importance to public life and identity, but they are not as well-established a part of the urban public landscape, and are readily made the objects of private commercial designs (Miller, 2002). Furthermore, about the only time of the year that Times Square is free of traffic is New Year's Eve, but this would not be the case with similar areas in European cities with pre-modern origins. You have to live in a city with places the type of *Place de la Concorde* to appreciate the way they function as an urban zero; a cipher that holds a place, an absent presence, a "nothing that is." I live near such a place in Copenhagen, *Rådhuspladsen*, and cross it regularly at all times of year, so I might as well begin with this "place."

Råd-hus-pladsen literally translates as "The Council House Place."[1] It is the open place in front of the Copenhagen city hall, where the city council, traditionally elected by the burghers of the city, meets to govern the city. The room made for by this empty place in front of the City Hall is a structural obverse to the space of the monumental building with its massive door and tower pointing into the heavens. This place is as empty and penetrable as the space of the hall is massive and impenetrable. Whereas the functions that take place within the building are clearly inscribed in the city's charter, the area in front of it is a kind of zero that holds a place for things to take place that may not be officially prescribed. A couple thus legally marries in a civil ceremony inside, but they become a publicly wedded couple when they emerge into the place in front, and are pelted with rice by well-wishers at the door. A civil wedding is called a *borgerligt bryllup* in Danish, literally a "burgher wedding," which is to say a wedding sanctioned by the laws of the burghers, rather than those of the church.

I use this quaint sounding term "burghers" because it refers to one of the three or four classic "estates" in which Western European society has historically been divided for political and legal purposes: the nobility, the clergy, the burghers and, sometimes, the free farmers. Much as *Rådhuspladsen* is identified with the burghers, there are other places scattered about the city that are similarly identifiable with other estates – a subject I will return to later. The burghers were the urban manufacturers and merchants of non-noble birth, who built the cities of Europe under the protection of the monarch's castles (burghs). The domain of the noble estate was in the countryside, but with the growth of cities the nobility built palaces in exclusive areas of the city, to be near the center of power. The estate of the farmers belonged exclusively to the countryside until the industrial revolution when many of the poorest farmers became urban workers. The industrial workers, under the influence of various socialist movements, sought to distinguish themselves as a class, separate from the burghers, and this helps explain the connotations of the term, especially when spelled in French, *bourgeois*. When used this way, it signifies the middle class (even the upper middle class), and hence the social values of that class, but it can also have a more neutral meaning as the non-noble citizenry.

There are other social groups that share the identity of an estate, though their status is less official than the traditional four listed above. The "fifth estate," of course, is that of the journalists, who are also sometimes linked to the "bohemians," the artists of the avant-garde. The professions, descendents of the medieval guilds, also share in the character of an estate, in the sense that they are groups with a particular social standing having particular duties and privileges – though these are threatened under the contemporary economic regime (Krause, 1996, 1998). Common to all is that they claim a particular quasi-legal status within the body politic that is analogous to the legal status of the official estates. Journalists, for example, claim the right and moral obligation to maintain the secrecy of their sources, even though this right rarely is written in law. The same applies to the scholars' claims to academic freedom, and the artists to "artistic license." Professionals, notably doctors, likewise subscribe to codes of moral behavior that carry a quasi-legal quality, defining them as a special group, with special rights and obligations within society.

The concept of an "estate" differs from that of an economic class because it is closely tied to the idea of a special legal status and representation within the body politic. Differing estates have been historically represented by different representative bodies, such as the upper and lower houses of parliament still found in some European countries.   The place of the burghers thus has symbolic importance as the locus of political influence of the non-nobles, the commoners. The idea of political representation identified with the differing estates, contrasts with that of modern democracy with its one-person one vote principle.   If the *res publica* of the estates is the shared area of the *place*, the symbolic space of modern democracy is the square voting booth, where the individual casts a vote in splendid isolation from the social and cultural matrices that otherwise define political life. Today one votes as an individual in a booth located within the space of the *Rådhus*, but one demonstrates for one's political views as a member of a social community on the place in front of the *Rådhus.*

The Nothing that is Something

On *Rådhuspladsen*, on a normal day, you will find a smattering of impermanent stalls where people hawk flowers or fruit.   There will also be several characteristic Danish *pølsevogne* (sausage wagons) – white metal contraptions on wheels that looks a bit like a small camper within which a *"pølsemand"* ("sausage man") sells his/her wares.   There is, finally, a fragment of a fountain (much of it was moved to make way for a six lane thoroughfare down the western side of *Rådhuspladsen*) and a couple of tubular constructions called *kiosks*.   Newspapers were traditionally sold at the quaint looking kiosks, but they have recently been modernized so that their main function seems to be to act as advertising space for newly privatized radio stations. With the possible exception of the *kiosks*, all structures and all signs of commercial activity are, in principle, impermanent, and can be removed at short notice.   *Rådhuspladsen* is in this respect like a zero.   The space within the 0 must remain empty, because the function of this place, like a cipher, is to hold a place – its significance lies in its emptiness, and because it is empty, things can take place here (figure 1).

Figure 1. Rådhuspladsen, Copenhagen, June 14, 2004, 5:59 p.m.. Facing north, in the opposite direction from the City Hall. From right to left: 1)a sausage wagon, 2)a kiosk, 3)an empty cafe, 4)the new bus station (filling entire background), source: author

Rådhuspladsen's year begins New Year's Eve when, as might be expected, the square fills with revelers and there is music and, of course, fireworks. On this occasion *Rådhuspladsen* is transformed into a place of celebration, as it is during other times of parades and carnival pageants, such as the homosexuals' parade that on their international day of struggle makes a passage from place to place through urban streets, turning the normal world upside down by breaking social barriers and generating ties to a liminal political landscape of community and place in the spirit of carnivalism (Bakhtin, 1984; Harrison, 1988; Turner, 1974). As the year progresses an enormous variety of planned and unplanned events might occur (figure 2).

This year saw opponents of the Iraqi war make several massive protests with anti-war speeches and songs from a temporary stage constructed in front of the city hall. There was also an ecological fair, complete with trees and lawns and a temporary soccer court in support of a youth initiative. A large poster presently announces an art project to turn the façade of the city hall into a place where citizens can express themselves. If the Danish soccer team should win the

European Cup, or a Danish cyclist win the *Tour de France*, then, as has happened in the past, there will also be a welcome home celebration on *Rådhuspladsen*.

Figure 2. Rådhuspladsen, Copenhagen, Sunday June 27, 2004, 6:03 p.m.. The day the Greek soccer team defeated the Danes in the Olymbic competition. The game is being shown on a wide screen at the base of City Hall, source: author

Transgressing Place

The importance of a place such as *Rådhuspladsen* lies not just in what can happen there, but also in the things that begin or end there, or that move through it. A protest demonstration often involves a march that ends at *Rådhuspladsen,* whereas May 1 workers' demonstrations might involve a march from *Rådhuspladsen* to a commons that was once on the outskirts of the city, but which is now a city park called *Fælledparken* (Commons park) (on Fælledparken see: Olwig, forthcoming). On May 14, 2004, when the crown prince (Frederik) married a commoner from Tasmania (Mary Donaldson) in the Copenhagen Cathedral, the couple went in an open horse drawn carriage from *Bispetorvet* (The Bishop's Place) along a route that took them first to *Rådhuspladsen*, where a large crowd welcomed them. People had been gathering there all day, beginning with a free breakfast served to thousands of people by the Danish farmers union[2].

Rådhuspladsen lies at one end of Copenhagen's narrow twisting and turning main drag, *Strøget*, along which people move (*stryger*) between places[3]. The old city historically faced on a multitude of places consisting of courtyards gated during the night, and a Swiss cheese of public places, such as *Kongens Nytorv,* the King's New Square, that lies at the opposite end of *Strøget* and is as noble as *Rådhuspladsen* is bourgeois. *Kongens Nytorv* dates from the 17th century and is surrounded by fancy noble palaces and the Royal Theater.   Beyond *Kongens Nytorv* lies a whole new Baroque era precinct, which has a grid of planned streets and is filled with palaces, including *Amalienborg*, the four royal palaces surrounding a public place. This is an area historically identified with royalty and the nobility, and it is still the fancy end of the old town. After passing through *Rådhuspladsen* the procession of the newly wed crown prince and princess moved down *Strøget* to *Kongens Nytorv*, and from there to *Amalienborg*, where a crowd had gathered to welcome the couple.

The movement of demonstrators marching from *Rådhuspladsen*, or the movement of the *prince* and his new princess procession, into and out of this place illustrates de Certeau's point that these places receive their meaning through the passages that transgress their bounds. It should be noted that de Certeau likes to bring out the full philological meaning of words by deliberately drawing upon their varying meanings. "Transgress" is thus used literally, as when referring to the wedding procession, to mean "to pass beyond or go over (a limit or boundary)," but when used to refer to demonstrators, the term begins to slide into the meaning of "to go beyond limits set or prescribed by (law or command): BREAK, VIOLATE" (Merriam-Webster, 2000: transgress). It is by transgression that these places are strung together through a movement that creates passages of meaning – as in the May 1 movement from *Rådhuspladsen* to the commons on the city's (former) outskirts, or the prince and princess's movement from *Bispetorvet* to *Rådhuspladsen, Kongens Nytorv* and finally *Amalienborg*.

The workers' transgression links the burghers' symbolic place to the symbolic place of the workers – the commons where the workers fought (literally) and won the right to celebrate May 1 about a century ago. The second transgression links places identified, historically, with the church, the burghers, the nobility and the royalty. Differing social groups know their historic places and transgress them in ways that transform the social and physical landscape, thereby "making history," as the TV announcer told the hundreds of thousands watching the live coverage of the royal procession from every possible angle, including that of vehicle moving parallel to the royal coach. History was indeed being made, as witnessed by the Danish national broadcasting service's pronouncement that that prince had found a "real princess."   They simply ignored the fact that Mary Donaldson was not actually a princess of noble blood, or even filthy rich (her father is a mathematics professor). History was made because the definition of nobility was being changed from one of birth to one of inner character, thereby marking a radical change in the makeup of contemporary Danish high society.

The place transgressing movements of the workers, in one class situation, and the prince and the new princess, in another, have the effect of emptying out spaces and filling them with new meaning. A commons becomes a memory laden site for the workers' movement, the place formed by the encircling buildings of *Amalienborg* is opened to the commoner well-wishers of a commoner princess, who feel a new sense of belonging within the closed precincts of a privileged elite.

Weaving Landscape

The procession of the prince and his princess did not end at the royal palace of *Amalienborg*. The crowd on the place in front of the palace waited patiently after the couple entered the palace to freshen up, because the couple was due to leave the palace again and proceed to their future residence at the royal palace of *Fredensborg*, north of the city. Here the citizens of *Fredensborg* town had erected a ceremonial entrance arch in honor of the newly weds. It was here that a gala wedding dinner was served not only for visiting royalty and dignitaries, but also for politicians from the parliament. A few days earlier a gala show had also been held at the house of parliament for the wedding guests. The carefully choreographed movement of the royal couple thus not only wove together different places within the city (the places of the clergy, the burghers, the nobility and the royalty), but also places connected to the larger political landscape of the country.

The transgressions that open up places do not necessarily go uncontested. Despite the air of high ceremony surrounding the progression of the royal couple from place to place within the city, the commentator for the Danish national broadcasting service could not resist making a highly critical comment about the desecration of *Rådhuspladsen* by a large shiny new glass modernistic bus terminal that party machine politicians had managed to have built on the margins of the square, despite massive public protests. This measure, forced through the City Council, broke unwritten taboos and led to the construction of what looked to be a permanent structure on *Rådhuspladsen*. The next transgression of rules occurred when a rightwing councilman gave permission to use the façade of the City Hall itself as screen for a giant laser advertisement for the premier of a Hollywood movie[4]. Places are not given and sacred, they are sites of struggle. A number of artists are thus presently developing a proposal to cover the façade of the building with a material that will allow the public (as opposed to Hollywood) to express their freedom of speech by painting texts and images on it.

The Nothingness of Place

Rådhuspladsen is a particularly good example of the nothingness of place because it lacks the qualities that generally have been seen to distinguish place from placelessness. To begin with, it does not have much of a history as a physical entity. It is a place that, about a century ago, was "moved," figuratively speaking, along with the city hall, from another location in the city. It was moved to a site that was made available by the tearing down of Copenhagen's ancient city wall. The importance of this place is thus not particularly related to a long history tied to a particular location. This is not to say, however, that the city hall square as *place* does not have a long history going back to the origins of the city. It is just that the history of this *place* happened, for the most part, at another *location* now called *Gammeltorv*, "Old Place." When the new city hall was built, the old city hall was torn down, creating space for a new place – *Nytorv*, "New Place." Physically, *Gammeltorv* and *Nytorv* form one contiguous open area, but they are nevertheless divided into two "zeroes" holding two places separated by a line going down, *Strøget.*

The essential quality of *Rådhuspladsen* as a place is not its physicality or location, but its quality as an ancient place (transferred from another location), held in place in front of the City Hall. Its physical qualities are not, unlike *Gammeltorv* and *Nytorv*, those that architects deify when writing of the genius loci of a place (Hunt, 2000) *Rådhuspladsen* would not be the functioning place that it is if its use depended upon a sense of place deriving from its physical qualities. Unlike the famous cozy sheltered and slightly hollowed place in front of Sienna's town hall, upon which *Rådhuspladsen* was patterned, *Rådhuspladsen* is cavernous, and consequently wind blown. Whereas Sienna's setting, under the windows of comely buildings sporting an array of cafés, attracts people like a magnet, the undistinguished buildings in a mishmash of styles that surround *Rådhuspladsen* create an uninviting space that leaves the chairs of its lone café largely empty most of the day, except for a few lost looking tourists.

The word was that the architect who designed the new bus terminal was caught off guard, along with the city fathers by the violence of the public reaction against the bus terminal, because *Rådhuspladsen* was seen to be the sort of space that one just rushes to get through. For the architect and the politicians *Rådhuspladsen* was simply *nothing*, an empty space waiting to be built upon, but they were clearly wrong. *Rådhuspladsen* is a "nothing that is," a zero patiently holding a place for things to happen – and when they happen a bus terminal is in the way. But this is not to say that the sense of place at *Rådhuspladsen* would not be improved by better architectural and planning practice. A reduction in traffic; the removal of the bus terminal; kiosks in a more appropriate style and the restoration of the fountain, would all strengthen the attractiveness of the place, but such measures do not, in and of themselves, create place.

LANDSCAPE IS PRACTICED LOCATION

After this examination of the place of place in the Danish political landscape, it is appropriate to return to de Certeau's complex conceptual realm. "Space is practiced place" according to the English translation of de Certeau's text (Certeau (de), 1984, p. 117). The French word *"lieu"* is here translated as *"place."* It would be clearer and more consistent, however, to translate *"lieu* with the word *location*, rather than with place. This is because, as noted earlier, de Certeau is specifically using "lieu" in conjunction with what the philosopher Merleau-Ponty termed "'geometrical' space ('a homogenous and isotropic spatiality')." *Lieu* is thereby: "the order (of whatever kind) in accord with which elements are distributed in relationships of coexistence. It thus excludes the possibility of two things being in the same location (*place*)." By *lieu* de Certeau thus means a location in geometric space, such as a coordinate in the geometric space of a map.

The term "space" as used in the above quoted phrase, "space is practiced place," is just as confusing as the use of *place* because the author does not mean space (*espace*) in the predominant English sense of the term, which is geometrical space, but rather in the sense used by Merleau-Ponty in reference to "anthropological space." In anthropological space "'space is existential'" and "'existence is spatial'," it expresses "'the same essential structure of our being as being situated in relationship to a milieu' – being situated by a desire, in dissociable from a 'direction of existence' and implanted in the space of a landscape" (Certeau (de), 1984, p. 117). Anthropological space is thus analogous to the milieu of a "landscape" – or *"paysage"* in the original French (Certeau (de), 1990, p. 174). *Milieu* is composed of the prefix *mi*, meaning "middle," and *lieu*. *Milieu* thus has a more substantial meaning than that of location; it means "surroundings," "medium" (as in chemistry) and "middle" (Deleuze & Guattari, 1988, p. xvii). Milieu, thus, is not a location in an abstract isotropic space, but a centered place that becomes a medium (potentially a happy medium or golden mean (*juste-milieu*)), wherein things take place.

De Certeau's concept of *milieu* bears something of the same relationship to *lieu*, as *place* bears to *location* in, for example, the above quoted passage from Tuan. It is through the practices by which locations are transgressed and "emptied out" that a place, or "room," is made for a "void" or "nowhere" that is liberated as "an area of free play" constituting the anthropological space of a "landscape" (*paysage*) (Certeau (de), 1984, pp. 104-106). In this way it gains more substance than the word location suggests and gains a history and meaning. This, of course, is literally what happens when *Rådhuspladsen* is transgressed by demonstrators, or revelers, emptying it of commercial activity, and transforming it into a place for free play where activities not normally tolerated in the space of the city can be accepted and sometimes even encouraged. *Rådhuspladsen* thereby ceases to be a *lieu*, and becomes a *milieu*, the environing "room" that holds a place for such activities to occur (on the distinction between room and space see Olwig, 2002a). These places, furthermore, are linked through the direction of the process of transgression into the larger anthropological space of a "landscape."

Room for things to take place

De Certeau's linking of place to a process of emptying out brings us back to the notion of zero, the cipher that holds a place. The concept of zero might appears to have originated in part in India, though when is hard to say, but it seems to have long been linked to the idea of place. The word "kha" was thus used by 500 AD to indicate a "place" in a series of numbers (as when we, today, delegate the place for tens as the second place to the left of the decimal – e.g. the place of the seven in 8<u>7</u>1.011), and this word later became one of the commonest Indian words for "zero" (Kaplan, 2000, pp. 42-43). The use of place was tied to the Indian way of thinking about the cosmos, and hence the godhead that created the cosmos.

The Indian concept of '*sunya* translates as "empty" but Indians did not have a concept of emptiness as an unqualified nothingness or void. As Kaplan explains: "Sunya isn't so much vacancy, then, as receptivity, a womb-like hollow ready to swell – and indeed it comes from the root' svi, meaning swelling. Its companion 'kha' derives from the verb 'to dig,' and so carries the sense of 'hole': something to be filled" (Kaplan, 2000, p. 59). This is, Kaplan argues, "the zero of the counting board: a column already there, but with no counters yet in it. This is the zero of the place-holder notation, having no value itself but giving value by its presence to other numerals" (Kaplan, 2000, pp. 59-60). A Greek variant of this kind of proto-zero notion of a nothing that is something is the concept of *chora*, as explicated by Plato in the *Timaeus*, where he calls it, as quoted by Kaplan, "'the Receptacle – as it were, the nurse – of all Becoming'" (Kaplan, 2000, p. 63). Another connotation of *chora* as receptacle, noted by the philosopher Jacques Derrida, is "imprint bearer" (Derrida, 1995, p. 93), which links it nicely to Kaplan's above mentioned idea that zero derives from the imprint left by counters in the sand, which then hold a place in the counting procedure.

The difficulty represented by the notion of zero in relation to place is illustrated by Aristotle's argument in the *Physics* that it might be thought that, in Aristotle's words, a "'void is a place where no body happens to be'." Aristotle then goes on to argue, however, that such a void cannot exist because, as Kaplan puts it, "with eternal things (and the elements that make up a body are eternal), there is no difference . . . between possible and actual Being – so all places are occupied (Kaplan, 2000, pp. 62-63). "Zero clashed," as Seife puts it:

> . . . with one of the central tenets of Western philosophy, a dictum whose roots were in the number-philosophy of Pythagoras and whose importance came from the paradoxes of Zeno. The whole Greek universe rested upon this pillar: there is no void. The Greek universe, created by Pythagoras, Aristotle, and Ptolemy, survived long after the collapse of Greek civilization. In that universe there is no such thing as nothing. There is no zero. Because of this, the West could not accept zero for nearly two millennia (Seife, 2000, p. 25).

The emptiness of *chora* was thus nearly unthinkable. The link between the notion of an emptiness from which all things are born and the godhead, which is linkable not only to Hindu mysticism, but also to the unsayable name of the Hebrew god Yahweh, made zero what Seife has termed, a "dangerous idea," preventing its emergence as an abstract mathematical concept until well into the Renaissance (Seife, 2000). Prior to the advent of zero, however, *chora* provided a name for this unnamable concept which, as the philosopher Martin Heidegger suggests, means: "that which abstracts itself from every particular, that which withdraws, and in such a way 'makes place' [*Platz macht*] for something else . . ." (quoted from An Introduction to Metaphysics, (Garden City, Doubleday, 1961, pp. 50-51)) by Derrida, 1995, p. 147, n.2).

Chora is often translated as "space," but this is not correct if the connotation of that word is "empty" or "absolute" space, implying extension or duration without the presence of a body or thing (Casey, 1997, pp. 34-35; Liddell & Scott, 1940, pp. 2016-2017, choros, chora). The Greek word for absolute or empty space was *kenos* (void) or *chaos* (Lukermann, 1967 (orig. 1961), pp. 55, 64), however, according to Kaplan, *chora*, like the Indian *sunya*, is not empty and infinite, but "carries the sense of a container ready to be filled," and it is characteristic of Plato that he fills it with numbers (Kaplan, 2000, pp. 63-64). Zero, as a "nothing that is," is not simply the absence of something, the empty void, but a concept which voids such a simple yes or no logic by denoting an absent presence. For Plato *Chora* was, in his own words, a "bastard concept" understood as "if in a dream" (Cornford, 1937; Derrida, 1995, p. 90) remaining amorphous and unknown until, according to the semiotician Julia Kristeva, it is given position and ordered according to the principles of geometry, which marks and "fixes it [chora] in place and reduces it" (Kristeva, 1984, pp. 25-27, 239-240). Something of this sort might have occurred when Renaissance city planners, inspired perhaps by the grid of Ptolemy's maps, transformed urban places into "squares" (Edgerton, 1975, pp. 122).

Chora as place, as choros

Chora, as the philosopher of place Edward Casey tells us, "is not just a formal condition of possibility. It is a *substantive* place-of-occupation," a "place providing" a "scene of emplacement" (Casey, 1997, p. 34). *Chora*, thus, was not just a philosophical concept, it also had a substantive meaning, and this meaning conveyed, as Derrida tells us, implications for "the discourse on *places* [places], notably political places" (Derrida, 1995, p. 104). This suggests that the meaning of *chora* might become clearer if one takes one's point of departure in the concrete realm of phenomena, as when the analysis of de Certeau's concepts took its point of departure in a concrete place, such as *Rådhuspladsen*. Philosophers usually use some variant of Plato's spelling such as *chora* or *khora*, but in more mundane discourse the spelling *choros* tends to be used, though both spellings refer to the same basic Greek concept (Liddell & Scott, 1940, p. 2016, chora, choros).

Choros, according to the Merriam-Webster dictionary, simply means "place" or "land" in ancient Greek (Merriam-Webster, 2000: choros). The standard

Oxford Greek-English Lexicon, "Liddell and Scott," elaborates upon this meaning, defining it as "a definite space, piece of ground,"place" and makes particular reference to "the lower world," defining it also as land or country" (Liddell & Scott, 1940, p. 2016, choros). The geographer, Fred Lukermann, defines *choros* as "literally meaning room," and he states that it "may safely be translated in context as area, region (regio), country (pays) or space/place – if in the sense of the boundary of an area. Choros technically means the boundary of the extension of some thing or things, it is the container or receptacle of a body" (Lukermann, 1967 (orig. 1961), p. 55). *Choros* had such related meanings as land (in the sense of country), region and place, as well as landscape, in the sense (discussed below) that it was an expression of the character of the land (Olwig 2001). A choros, for the Greeks, was first and foremost a polis, with its political and cultural focus on the things that took place, or occurred, in the roomlike place of the *agora* where an aggregate of people assembled (also called *agora*) in order to carry out a discourse (also called *agora*) (Hénaff & Stong, 2001, p. 45). The *agora* was the ancient Greek equivalent of *Rådhuspladsen*.

Plato's notion of *chora* as a feminine receptacle that gives birth to the world fits well into the ancient "sexual cosmology" of the sky (the locus of Plato's ideal cosmic archetypal geometric forms that are expressive of paradigmatic ideas) as a male force impregnating a female earth, or Gaia (Olwig, 1993). In this case his *chora* might well have been inspired by autochthonous liminal sacred places like the *agora*, and their association with rites and festivities linked to the fertile character of the landscape or choros (Coulanges, n.d. (orig. 1864); Morris, 1991; Olwig, 1993; Polignac, 1995). *Chora/choros* defied Plato's geometric rationality by defining itself as a nothing that is something, for how could one grasp the archetypal form of something that was nothing, and therefore had no form? How could one deal with a liminal space that defined itself as being qualitatively separate from other space (Eliade, 1959)? The only solution was to define *chora* as a kind of conceptual antithesis to the ideal forms of platonic rationality. The second century Greek Platonist, astronomer, astrologer, mathematician and geographer, Claudius Ptolemy, wrestled with similar issues when he sought to map *choros* using the tools of geometry and mathematics, based on a study of the cosmos.

Mapping choros

Choros is the root of chorography, one of the three branches of geographical science outlined by Ptolemy. Chorography was concerned with the delineation and regional description of the lands of the earth as places. Though Plato made *chora* a key concept in his cosmology, he was also clearly flummoxed by the idea of a nothing holding a place for a something, calling it a "bastard" concept. Ptolemy, likewise, needed a concept of *choros* for his geography, but he was not able to deal with it within his mathematical system, so he relegated it to the realm of art and the qualitative. Geography itself, for Ptolemy, was a mathematical science concerned essentially with the mapping of locations as position points upon the graticule of a globe and the study of their spatial relations. Geography, in Lukermann's words, was concerned with "processes of world-wide pervasiveness," or to use a more contemporary phrase, it was concerned with the *global*. Topography was defined as" the order of discrete units one to the other" as plotted on the graticule of the map (Lukermann, 1967 (orig. 1961), p. 55). Chorography, on the other hand, was concerned with the hollow spaces of the regions within which lands are shaped as places, a clearly important subject, but one which defied mathematical treatment. As Ptolemy explained:

> Geography looks at the position rather than the quality, noting the relation of distances everywhere, and emulating the art of painting only in some of its major descriptions. Chorography needs an artist, and no one presents it rightly unless he is an artist. Geography does not call for the same requirements, as any one, by means of lines and plain notations, can fix positions and draw general outlines. Moreover Chorography does not have need of mathematics, which is an important part of Geography (Ptolemy, 1991, p. 26>)

Ptolemy hereby brings us back to the basic issues, broached by Relph and Tuan, concerning the distinction between the analysis of locations in a gridded geometric space and the understanding of place; issues that were also of concern to de Certeau. For de Certeau, however, place is not just a kind of antithesis of location in space, it is also an emptiness that can be "made habitable" as a milieu woven into a "landscape" (*paysage*). If the *agora* is a kind of Greek equivalent to urban places like *Rådhuspladsen*, then the *choros* is roughly equivalent to the political landscape woven together through the transgression of, and movement between, such places. However, this use of the concept of landscape differs, from the notion of landscape as visual scenery that British scholars tend to identity with the concept (Cosgrove, 1984; Duncan, 1990), it can thus be useful to go into a bit more depth concerning the use of landscape/*paysage* made here.

LANDSCAPE AND *PAYSAGE*

Paysage appends the suffix -*age* to pays in much the way as -*schaft* is appended to *Land* in the Germanic languages, or -*ship* to *town* in English. *Pays* carries essentially the same connotations of areal community and people as country and land in English. The equivalent Italian terms, *paése* and *paesàggio*, likewise carry the same meaning (Battisti & Alessio, 1975: paése, paesàggio; Gamillscheg, 1969: pays, paysage; Robert, 1980: pays, paysage). The root of landscape and *paysage* is land/*pays*, in the sense of a place or country of dwelling, not a land surface. It is a political landscape rather than a panoramic landscape. The suffix *scape* is equivalent to –*ship*, meaning, among other things, "state" in the sense of condition, or quality – a meaning shared by the French suffix – *age* (Merriam-Webster, 2000:

ship, age). The word friend*ship* thus refers to more than the act of being a friend, but to the state or quality of being a friend. It makes sense, therefore, to term a painting of places in a land or *pays* a land*scape* painting, or a painting of a *paysage*, because it seeks to capture the state or quality of a land/*pays*, or, following Ptolemy (whose geography appears to have exerted a profound influence on Renaissance art), of a *choros* (Edgerton, 1975, p. 120). Landscape in this sense is primarily a place, or an assemblage of places, not a visual scene or panorama (on the evolving meaning of landscape see Olwig, 2002b). This panoramic notion of landscape does play a role in de Certeau's analysis. It is not, however, a landscape woven together of places, but the view seen from locations such as that of the tower of the World Trade Center.

The WTC, according to de Certeau, is "the most monumental figure of Western urban development" (Certeau (de), 1984, p. 93). Quoting Michel Foucault, de Certeau argues that the WTC is the materialization of a scopic utopia that the artists of earlier times could only imagine as a vision of how the world would look if seen from the perspective of the gods with the "'all seeing power'" of a "celestial eye" (Certeau (de), 1984, p. 92)[5]. The WTC is the premier icon of what de Certeau calls "the Concept city," emerging in the Renaissance as an expression of its political model, "Hobbes' State" (the centralized state advocated in Thomas Hobbes' *Leviathan*) (Certeau (de), 1984, pp. 94-95; Hobbes, 1991 (orig. 1651)). This conceptual city was superimposed upon the fact of existing cities as part of the process by which the mercantilist centralized state was established in Europe.

The view from the top of the WTC embodies the ideal landscape of the central state as expressed through the graphic representation of what de Certeau terms "the space planner urbanist, city planner or cartographer." "The panorama-city" created by these mappers is fundamentally "a 'theoretical' (that is, visual) simulacrum, in short a picture" (Certeau (de), 1984, pp. 92-93)[6]. The ability to visualize the world as panoramic scenery developed during the Renaissance on the basis of the techniques of surveying and cartography that were inspired in some measure by the rediscovery of Ptolemy's cartographic techniques (Edgerton, 1975, 1987). It creates, in de Certeau's words, its own "rationally" organized space and its own "synchronically" ordered "nowhen" of progress in time that is privileged over history (Certeau (de), 1984, p. 94).

The map or plan is projected on the coordinates of a graticule, or geometric grid that then provides the basis for the spatial organization of the city and the regionalization of the landscape as territory in a gridded, striated, space. The surveying, mapping, and division of land into quadratic spaces facilitates their measurement into quadratic properties that can be valued and weighted according to the equal measure of a given square unit as viewed on a map or plan. It thus scales and reduces qualitatively different areas, such as open city places, building sites, meadowlands, forestlands, croplands, allodial lands and crown lands, to a neutral commodity that can be sold, "fair and square," by the common denominator of a square unit.

Though the word *scale*, when applied to the device utilized for weighing, derives from a different etymology than the *scale* of a map, it has the similar function of facilitating the reduction of qualitatively different things to a quantitative common denominator. Both the scales used to weigh and the scale used to measure and map thus greatly facilitate the growth of a mercantile economy by creating equal units that may, in turn, be reduced to commodities sold by the common denominator of money. By being made into a commodity units of weight thus become common denominators in a trade economy, and hence synonymous, so that the verb *scale* comes to incorporate both units of weight and spatial measures: "to make or to lay out so as to be of exact weight, quantity, or dimensions" (Merriam-Webster, 1968: scale).

Because maps are made to different scales they can also be used to create the illusion of an homology between units that appear as the same size on a sheet of paper, or globe. In this way, for example, the island of Britain, floating in the sea, could be compared to a world floating in the heavens, or a city might be compared to an ant colony or bee's nest, even though these phenomena have little in common beyond scalar space. Altering scale creates an effect similar to the perspective on the world that one gains when standing at the top of the WTC, by which a qualitatively large, and hence incomprehensible, panoramic landscape, is reduced to what appears to be a manageable and comprehensive unit.

The geometries of the space planner urbanist, city planner or cartographer generate, in de Certeau's words, the common denominator of "a *universal* and anonymous *subject* which is the city itself: it gradually becomes possible to attribute to it, as to its political model, Hobbes' State, all the functions and predicates that were previously scattered and assigned to many different real subjects – groups, associations, or individuals. The city, like a proper name, thus provides a way of conceiving and constructing space on the basis of a finite number of stable, isolatable, and interconnected properties" (Certeau, 1984, p. 94). The same analysis could be applied to the contemporary use of the global, which has likewise become a universal and anonymous subject constructed on the basis of a finite number of stable, isolatable, and interconnected properties. This subject is conceived in a scalar space in which it is possible to trace an infinity of lines of longitude and latitude, allowing scalar movement not only from the global to the local, but also to the microscopic and the infinite. The illusion is thus generated that somehow all phenomena can be scaled in space, thereby implying that there is a common denominator between the local location and the global. Such a common denominator, of course, obliterates difference, including the difference between space and room – a key element in both fascist and modernist ideology (Olwig, 2002a). The space of the city hall tower's linear mark thus becomes coequal to the room of the zero that lies before it, thus obliterating the distinction between a place that is to be held empty and a building site, thus justifying the architect's building of the bus terminal on a portion of its space.

The creation of perspective representation, according to Charles Seife, was an expression of the breakthrough of the abstract mathematical concept of zero (Seife,

2000, pp. 84-87). The perspective drawing, as Seife points out, is constructed within a space focused on lines drawn from the eye to a point of infinity that is the focus of the picture. Neither the focal point of the central point perspective picture, nor the framed space of the drawing, however, is the emptiness of a zero. It is rather the infinite space of the grid, upon which things are located. The painting does represent the breakthrough of a new world view that is willing to accept the infinity of space (Nicholson, 1959), and this also, no doubt, facilitated a willingness to accept the nothingness of zero, and hence its use in abstract mathematics. But because of its very abstractness, in relation to the concrete, this zero looses its integral relation to place. It still holds a place, but it now seems farfetched to suggest, as I have suggested, that there is a connection between the kind of place held by zero in modern mathematics, and the kind of place held by *Rådhuspladsen*. Zero now holds a place, but only in an abstract mathematical sense, divorced from the exigencies of places in the material world.

The space of the landscape panorama, like that of the map, belongs to the abstract realm of an infinite isotrophic space within which things are located. It reduces place to a geometric structure of locations. This is the space that de Certeau experienced from a location on the 110th floor of the World Trade Center. This is the location that must be transgressed if it is to be transformed into a place in a landscape.

GROUND ZERO: THE TRANSGRESSION OF LOCATION

De Certeau's vision of Manhattan as seen from the World Trade Center is filled with disturbingly cataclysmic metaphors characterizing a concept city that is decaying. At the same time as the concept city decays, however, the pedestrian urban life, which preceded the concept city, continues apace amongst the masses that throng beneath the tower. New York is a city "composed of paroxysmal places" in a "universe that is constantly exploding." The viewer, like a modern day Icarus, is carried up and out of the city's grasp – and risks a similar "Icarian fall." The unease attendant to ascending the tower is compounded, for de Certeau, by the sphinx-like enigmatic message of a poster on the 110th floor *"It's hard to be down when you're up,"* which for an instant transforms the pedestrian into a "visionary" (Certeau (de), 1984, pp. 91-95). The violent transgression by which terrorists downed the World Trade Center to "Ground Zero" is foreign to de Certeau's more benign notion of the "emptying out" process, yet it nevertheless represents an extreme example of the transgressive emptying out process by which a marked location is transformed into an open place.

Ground Zero is an empty place, where the icon of the World Trade Center once was located, and it has become "an area of free play," an anthropological space which has become central to the identity of New York as the landscape of a polity. The significance of emptiness is brought out in the writer, Adam Gopnik's, interview with the urban theorist Jane Jacobs. Gopnik writes:

Jacobs has closely followed the Ground Zero plans and debates, and she thinks that the right thing to do is not to do anything right away. "The significance of that site now is that we don't know what its significance is," she said. "We'll known in fifteen or twenty years. " She also thinks that it might be a good idea not to "restore" the street grid at the Ground Zero site but to break it decisively. "I was at a school in Connecticut where the architects watched paths that the children made in the snow all winter, and then when spring came they made those the gravel paths across the green. Why not do the same thing here?" (Gopnik, 2004, p. 34).

The WTC was a beacon for global capital, providing office space for world traders with homes all over the globe and dwellings typically in suburbs, such as Westchester County, north of the city. The building's fall carried many of these world traders to their death, but it was the indigenous New York firemen, who were buried under the building, not the stockbrokers, that became the primary object of public mourning. Whereas the tower was a monumental location belonging to the globe, the hallowed hollow of Ground Zero is a place belonging to the political landscape of New York City, weaving together the towers and concrete of Manhattan with the homes and lawns of places like Staten Island, where many of the firemen lived. The importance of Ground Zero, however, goes beyond New York, linking, for example, the firemen from all over the world who expressed their solidarity with their New York brethren in ways that often involved self-sacrifice.

The panorama seen from the 110th floor of the WTC is a construction of the map and globe, it is a realm constituted by networks of locations in space. It is an expression of a global vision, and imperial global domination going back to ancient Greece, and revived in the Renaissance (Cosgrove, 2001). The world centering on Ground Zero is equally large, but it is a landscape of places woven into a complex multitude of political landscapes making up an anthropological space that, like zero, may seem like nothing, but which actually constitute a nothing that is – an absent presence – and it has the potential to undermine the global hegemony represented by locations such as that of the World Trade Center.

Department of Landscape Planning, Swedish Life-Sciences University, Alnarp, Sweden

NOTES

[1] The definite article (spelled *en* or *et*) is placed after the word it modifies, so *Rådhuspladsen* means the *Rådhusplads*. Since the name in Danish is *Rådhuspladsen* I have decided to spell it in the Danish way, and drop the English definite article.

[2] The word *torv* is a Nordic word meaning "open place." It is related to similar words in Old Russian and perhaps Mongolian that refer to market place trade (Politiken, 2002: torv).

[3] *Strøget* derives from a word meaning something like "to glide."

[4] Copenhagen's "magistrate" system of political organization apportions authority over different branches of city government (e.g. schools, planning etc.) to "sub-mayors" from the multiple political

parties represented on the council. This means that even though the left leaning Social Demnocratic party may have the majority, with the Lord Mayor coming from that party, a member of another party will have control over specific departments, such as the one that gave permission to use the wall of city hall for an advertisement.

[5] The history of the idea of the celestial eye has been charted in (Cosgrove, 2001).

[6] The word "theoretical" derives from the "Late Latin *theoreticus*, from Greek *theoretikos*, from *theoretos* (verbal form of *theorein* to look at, behold, contemplate, consider)." De Certeau compares Manhattan to a "stage" (Certeau, 1984, p. 91), and it should be noted that the word "theater" has a similar etymology "Middle English theatre, from Middle French, from Latin *theatrum*, from Greek *theatron*, from *theasthai* to see, view (from *thea* action of seeing, sight) + -tron, suffix denoting means, instrument, or place; akin to Greek thauma wonder, miracle" (Merriam-Webster, 1968: theory, theater).

REFERENCES

Bakhtin, M. (1984). *Rabelais and His World*. Bloomington: Indiana University Press.

Battisti, C., & Alessio, G. (1975). *Dizionario Etimologico Italiano* (Vol. IV). Firenze: Instituto Di Glottologia, G. Barbèra.

Casey, E. S. (1997). *The Fate of Place: A Philosophical History*. Berkeley: University of California Press.

Certeau (de), M. d. (1984). *The Practice of Everyday Life* (S. Rendall, Trans.). Berkeley: University of California Press.

Certeau (de), M. d. (1990). *L'invention du quotidien: Arts de faire* (New Edition Vol. I). Paris: Gallimard.

Cornford, F. M. (1937). *Plato' Cosmology: The Timaeus of Plato, translated with a running commentary*. London: Routledge & Kegan Paul.

Cosgrove, D. E. (1984). *Social Formation and Symbolic Landscape*. London: Croom Helm.

Cosgrove, D. E. (2001). *Apollo's eye: A Cartographic Genealogy of the Earth in the Western Imagination*. Baltimore: The Johns Hopkins University Press.

Coulanges, N. D. F. d. (n.d. (orig. 1864)). *The Ancient City* (W. Small, Trans.). Garden City, N.Y.: Doubleday Anchor.

Deleuze, G., & Guattari, F. (1988). *A Thousand Plateaus: Capitalism and Schizophrenia*. London: Continuum.

Derrida, J. (1995). Khora (I. McLeod, Trans.). In J. Derrida, *On the Name* (pp. 88-127). Stanford: Stanford University Press.

Duncan, J. S. (1990). *The City as Text: the Politics of Landscape Interpretation in the Kandyan Kingdom*. Cambridge: Cambridge University Press.

Edgerton, S. (1975). *The Renaissance Rediscovery of Linear Perspective*. New York: Basic Books.

Edgerton, S. (1987). From Mental Matrix to Mappa mundi to Christian Empire: The Heritage of Ptolemaic Cartography in the Renaissance. In D. Woodward (Ed.), *Art and Cartography: Six Historical Essays* (pp. 10-50). Chicago: University of Chicago Press.

Eliade, M. (1959). *The Sacred and the Profane*. New York: Harcourt, Brace & World.

Gamillscheg, E. (1969). *Etymologisches Wörterbuch Der Französischen Sprache* (2 ed.). Heidelberg: Winter.

Gopnik, A. (2004, May 17). Urban Studies: Cities and Songs. *The New Yorker,* 30-31.

Harrison, M. (1988). Symbolism, 'Ritualism' and the Location of Crowds in Early nNneteenth-century English Towns. In S. Daniels and D. Cosgrove (Eds.), *The Iconography of Landscape: Essays on the symbolic representation, design and use of past environments* (pp. 194-213). Cambridge: Cambridge University Press.

Hénaff, M., & Stong, T. B. (Eds.). (2001). *Public Space and Democracy*. Minneapolis: University of Minnesota Press.

Hobbes, T. (1991 (orig. 1651)). *Leviathan (orig. title Leviathan, or The Matter, Forme, & Power of a Common-wealth ECCLESIASTICALL AND CIVILL)*. Cambridge: Cambridge University Press.

Hunt, J. D. (2000). *Greater Perfections: The Practice of Garden Theory*. London: Thames and Hudson.

Johnston, R. J. e. a. (Eds.). (2000). *The Dictionary of Human Geography* (4 ed.). Oxford: Blackwell.

Kaplan, R. (2000). *The Nothing that Is: A Natural History of Zero*. Oxford: Oxford University Press.

Krause, E. A. (1996). *Death of the Guilds: Professions, States, and Capitalism*. New Haven: Yale.

Krause, E. A. (1998). The Politics of Professional Expertise. In L. O. Vittorio OIgiati, Mike Saks (Eds.), *Professions, Identity, and Order in Comparative Perspective*. Onati: The International Institute for the Sociology of Law.

Kristeva, J. (1984). *Revolution in Poetic Language* (M. Waller, Trans.). New York: Columbia University Press.

Liddell, H. G., & Scott, R. (1940). *A Greek-English Lexicon* (A New Edition revised by Henry Stuart Jones and Roderick McKenzie Eds.). Oxford: Clarendon Press.

Lukermann, F. (1967 (orig. 1961)). The Concept of Location in Classical Geography. In L. M. Sommers and F. E. Dohrs (Eds.), *Introduction to Geography: Selected Readings* (pp. 54-74). New York: Thomas Y. Crowell Company.

Merriam-Webster. (1968). *Webster's Third New International Dictionary of the English Language, Unabridged*. Springfield, Mass.: G. & C. Merriam.

Merriam-Webster. (2000). *Webster's Third New International Dictionary of the English Language, Unabridged – Electronic edition*. Springfield, Mass.: Merriam-Webster.

Miller, K. (2002). Condemning the public: Design and New York's new 42nd Street. *GeoJournal, 58*, pp. 139-148.

Morris, I. (1991). The Early Polis as City and State. In A. Wallace-Hadrill. John Rich (Eds.), *City and Country in the Ancient World* (pp. 24-57). London: Routledge.

Nicholson, M. H. (1959). *Mountain Gloom and Mountain Glory: The Development of the Aesthetics of the Infinite*. New York: W.W. Norton.

Olwig, K. R. (1993). Sexual Cosmology: Nation and Landscape at the Conceptual Interstices of Nature and Culture, or: What does Landscape Really Mean? In B. Bender (Ed.), *Landscape: Politics and Perspectives* (pp. 307-343). Oxford: Berg.

Olwig, K. R. (2001). Landscape as a Contested Topos of Place, Community and Self. In S. H. Paul C. Adams, Karen E. Till (Eds.), *Textures of Place: Exploring Humanist Geographies* (pp. 95-117). Minneapolis: The University of Minnesota Press.

Olwig, K. R. (2002a). The Duplicity of Space: Germanic 'Raum' and Swedish 'Rum' in English Language Geographical Discourse. *Geografiska Annaler, 84B*(1), pp. 29-45.

Olwig, K. R. (2002b). *Landscape, Nature and the Body Politic: From Britain's Renaissance to America's New World*. Madison: University of Wisconsin Press.

Olwig, K. R. (forthcoming). Representation and Alienation in the Political Land-scape. *Cultural Geographies*.

Polignac, F. (1995). *Cults, Territory and the Origins of the Greek City-State* (J. Lloyd, Trans.). Chicago: University of Chicago Press.

Politiken. (2002). *Politikens Store Ordbog*. Copenhagen: Politikens Forlag.

Proux, A. (2001, January 1). Influences: Big Skies, Empty Places. *The New Yorker, 139*.

Relph, E. (1976). *Place and Placelessness*. London: Pion.

Robert, P. (1980). *Dictionnaire Alphabétique et Analogique de la Langue Française* (Vol. 5). Paris: Le Robert.

Seife, C. (2000). *Zero: The Biography of a Dangerous Idea*. New York: Viking.

Tuan, Y.-F. (1974). Space and place: humanistic perspective. *Progress in Geography, 6*, pp. 211-252.

Turner, V. (1974). *Dramas, Fields and Metaphors: Symbolic Action in Human Society*. Ithica: Cornell University Press.

JUSSI S. JAUHIAINEN

IN POST-POSTMODERN TECHNOLOGISED LANDSCAPES

Technobus, source J.S. Jauhiainen

INTRODUCTION

In one commercial advertisement from the 1970s on Finnish television a family had gathered together around the table. They started to eat breakfast when suddenly the young boy asked his father: "Dad, what did we use before there was Flora". The whole family looked surprised at the boy. After a short but definitive silence the father replied: "What – there was no Flora?". Flora was a label for a new margarine, a particular technological innovation launched to compete with butter. In the past months this advertisement has been shown again on television.

T.S. Terkenli & A-M. d'Hauteserre (eds.), Landscapes of a New Cultural Economy of Space, 193-211.
© *2006 Springer. Printed in the Netherlands.*

Let me rephrase this dialogue for the early 2020s. "Ma, what did we do before there were personal computers" – "Dad, what did we do before there were mobile phones" I'm sure that the face of the young parents would be more astonished than any copywriter could write or any actor could play for a television advertisement – no personal computers in the office, no mobile phones available, no global information and communication technology…?

One aspect of the new cultural economy in the 1990s has been the transformation of everyday landscapes. Cities, regions and countries that are connected to economic flows and consumption-oriented global capitalism are said to suffer from "Macdonaldisation", "Disneyworldisation", "Microsoftisation" and "Nokiasation". In the past few years modern information and communication technologies (ICT) have become common in our everyday life and landscapes. They have today almost a self-evident and unquestionable presence. The new generations of mobile phones are transforming into complex mobile communicators. They connect us continuously through networks and flows to other landscapes at different spatial distances and shape our time-space practices, integrating the realms of public and private, home, work and leisure. To get along in everyday life one needs ICT and with them life becomes "faster and more intensive". People carry taken-for-granted technology and have access to and are being accessed by global information and communication flows.

The future city is no more a place of escape, to get lost within and to step aside. The body, the mind and the environment are particularly mixed and melted into a simultaneously present and absent here/t/here technologised landscape surrounding, intertwining and penetrating us. To define 'technologised landscape' I lean on Olsson (2003, p. 300) who states: "The reason is that a landscape is not as simple as it might first appear, not limited to what meets the eye but equally present in the mind. Put differently, landscapes are in neither body nor mind, but in the conjunction of the two." In a 'technologised landscape' the traditional ontologically neat geographical division between body and mind, here and there, and place and space vanishes.

Technologised landscapes and their signs are encompassing all worlds into one without negotiation. We cannot escape from technologised landscapes in which use and exchange values are intertwined and melting into one commodity. Economic globalism leads to this commodification of landscape or deworldment processes. It produces a certain inauthenticity in landscapes by facilitating the penetration and materialisation of globally circulating economic flows into locally built environments. It transforms the material environment into a 'hyperspace' that covers the spatio-temporal differences (here / there / past / present / future) in landscape (Jameson, 1991).-These "inauthentic" landscapes become the taken-for-granted everyday landscapes of people. The issue is not only or mostly about the authoritarian surveillance by CCTVs and the control of electronic personified databases but about more subtle modes of attraction, inclusion and exclusion. ICT remind and attract us towards "correct" spatial performances in consumption, leisure and work.

A technologised landscape is a global phenomenon but it is not yet omnipresent. According to current statistics, in the OECD countries there are more users of mobile phones than of traditional phones. In the World there are eight countries where the share of mobile phone subscribers is over 90 per cent of the population (in Finland 90.1 per cent). In ten countries the majority of population are users of Internet (in Finland 50.9 per cent) and in eight countries there are fewer than two persons per one personal computer (in Finland 2.3 persons) (ITU, 2004).

Still today, the vast majority of human beings have never used modern ICT, mobile phones, personal computers or the Internet or actually seen buildings in which the ICT devices are produced. Nevertheless, in 2004, about one fourth of the global population has a mobile phone. Every day more than a million mobile phones are sold and the number of mobile phone subscribers increases by half a million. Taking into account the development in past years by 2006-07 there will be over two billion mobile phone subscribers (ITU, 2004). Relatively soon the whole globe and its population will become an ICT network in which the time-space presence / absence will be very different than just a decade ago.

THE STUDY OF TECHNOLOGISED LANDSCAPES

How to approach the notion of landscape in the era of technologised landscapes? From ontological and methodological perspectives I indicated earlier that a major challenge for the latter 20th century landscape studies was how to overcome the binary between morphologists and representationalists (Jauhiainen, 2003). There is a clear division between those whose aim is to describe and divide physical and material landscapes into internally homogeneous units and those who study the landscape in the minds of people or see landscape representing something beyond its physical appearance (Jauhiainen, 2003). For the early 21st century Mitchell (2003, p. 790) claims the need for both a [general] theory of landscape and theories of capital circulation, of race and gender, and of geopolitics and power [regarding landscape]. Henderson (2003, p. 336) argues that the cultural, social, political and economic ought to be folded into landscape to understand why the cultural, social, political and economic matter.

According to Matless (1998, p. 12), the question of what landscape is can always be subsumed in the question of how it works – as a vehicle of social and self identity, as a site for the claiming of cultural authority, as the generator of profit, as a site for the reconstruction of a [particular] citizenship (Mitchell, 2001, p. 277). Instead of a detailed conceptualisation of landscape or theory about technology in this chapter I approach broader questions such as who are we, what is this time, what is characteristic of our epoch and in which landscape we live the processes of deworldment, cultural transformation and the emergence of "inauthentic" landscapes. For Backhaus & Murungi (2002, p. 10), landscape is the earth's textuality, a geographic organisation of the once natural and social life-world in which the landscape as a text is a hermeneutic between humans and the meaning-sedimented natural and social life-world. However, in the technologised landscape the notion of landscape is more flexible in its connection with the natural and social. The hybrid combination of human and technology continuously modifies the landscape: there is neither time nor place for "sedimentation". The technologised landscape is a worknet – the work, the movement, the flow, and the changes are stressed (Latour, 2004).

The ICT are imprinted on the landscapes almost everywhere dissolving visually "authentic" landscapes. Broadly distributed and instantaneously replicated collective junk mail, banners, icons, electronic advertisement, homepages and Internet information through physically similar ICT devices are part of the contemporary collective technologised landscapes. Nevertheless, one should carefully examine these "inauthentic" landscapes in their particular contexts. The names, the signs, and the symbols in the technologised landscape are the same but there are (still) variations in the socio-spatial daily practices. For example, in the Philippines the average daily number of text-messages per mobile phone user is counted in tens and in many countries it is less than one. Despite the invisibility of these information flows in the landscape these text-messages leave traces and information both materially and immaterially. There is a particular geodiversity in technologised landscapes, i.e. the outcomes of ICT-mediated economic globalism are not exactly the same everywhere but this "sensitivity" towards local landscapes might be just avariation in global capitalism to ensure that this globalism will be present in every landscape.

Methodologically the approach in this chapter is a contemporary diagnosis about contemporaneity, called *Zeitdiagnose* in the German-speaking social studies. The methodological rigour in *Zeitdiagnose* ("diagnosis of the era") is the belief in a subjective synthesis in which grasping, understanding, vision and openly committed viewpoints are important (Noro, 2004). These open possibilities to say and write something right now about the passing moments. *Zeitdiagnose* helps to integrate the moment, general theoretically informed observations and the empirical context. The uncertainty, the committed tendency to question the legitimacy of scientific methods and the truth as well as the recognition of unknown possibilities for action are important. They permit one to see how a seemingly universal technologised landscape becomes contextualised in local micro-settings. Zeitdiagnose as methodological standpoint also allows the use of the findings in other contexts and to question alternative possibilities with a rigorous critique of the everyday social realities. The experience of deworldment is much more than can be seen on the façades of the material environment. The performances of everyday practices connect local specificities and practices with other contexts so the glittering façades do not mirror ourselves exactly. In this way Zeitdiagnose also helps to understand how deworldment is specific in every context and for every person.

The empirical study in this article is deliberately narrowed into "a personal narrative of my everyday environment" in Oulu and its surroundings, a Finnish town where I live and work. The chapter is an attempt to study the technologised everyday landscapes from a post-postmodern standpoint. The idea is to consider what it is to be "in the post-postmodern" when everydayness is technologised to an extent that one hardly pays attention to it or is distanced from it. The post-postmodern is the era in which the cyborgs are "naturalised". The cyborg was a feature of scientific fiction in the 1960s and a topic of intellectual debate in the 1980s-90s (Haraway, 1991). In the 2000s, the combination of genome-, bio-, information and communication technologies at the nanoscale blurs further the boundaries between people and machines. In the "Western world" we all are becoming cyborgs through everyday actor-networks between human beings, technology and the environment: fused into the technologised landscapes (Latour, 1993). ICT can be thought of as extra organs growing into existence, rather than as something outside the compass of the human body (Amin & Thrift 2002, p. 78). We are connected to and part of these technologised landscapes through the ICT mediators: mobile phones and communicators, personal and lap top computers, materiality of ICT-related enterprises in our near environment, and the electro-magnetic flows that surround us.

According to Amin & Thrift (2002, p. 52), contemporary urban settlements are assemblages of more or less distanced economic relations with different intensities at different locations. Globally, my case town Oulu is small and peripheral in the north. The town has 126,000 inhabitants (200,000 when including the surroundings) and it is located at 65°01' northern latitude and 25°28' eastern longitude. Nevertheless, during the 1990s Oulu gained national and international recognition as a significant technology centre and today it is globally among the leading sites in the development of mobile technologies. Among hundreds of high-tech enterprises there is a large research and development centre for Nokia, the world's largest mobile phone producer. In the area over 6,000 persons work in high technology. Furthermore, at the University of Oulu there are about 5,000 students in the Faculty of Technology.

Oulu is a curious "frozen Silicon Valley" located at the northern fringe of the ecumene. The town is slightly ahead in many, often naïve practical activities regarding mobile technologies. For example, mobile phones and communicators can be used to purchase a fishing license, to order food in a restaurant, to pay a parking fee, to be reminded of the weather changes, to buy a ticket to recreational activities and to be directed to different consumption possibilities within the town centre. Some inhabitants have even volunteered for free to test the new mobile ICT applications. Furthermore, in the Finnish media and political discourses the "technological miracle of Oulu" is a widely used narrative of success. Therefore Oulu is well suited to the method of *Zeitdiagnose* to consider what it is to be exposed to technologised landscapes.

The rigid separation of space from place and the clear-cut layering of scales has been criticised in recent geographical studies (Smith, 1993; Brenner, 2001; Mitchell, 2003; Massey, 2004). To recognise such critiques I use flexible 'scale-jumping' as an interpretative frame to indicate simultaneous connections in the technologised landscapes and in their politics of scale. I approach technologised landscapes and scale through an everyday bus trip from the airport of Oulu to the technology centre and to my office. What is important is to position oneself in the technologised landscape. The approach is loosely connected to Bruno Latour's (1993; 1996; 1999; 2004) ideas about ourselves as human–non-human hybrids being surrounded by and taking part in various human-technology actor-networks. It is the personal and the contextual in technologisation that can be used as a possibility to question the threads of linear understanding of deworldment – seemingly similar but still different kinds of activities and outcomes in the material, experienced and symbolic technologised environment.

The case study starts at the airport of Oulu. There one finds a bus that travels a seventeen kilometres long road through three municipalities to the university and to the technology park of Oulu. I follow this road almost weekly and always when I leave and come to Oulu. Most often I use public transport, bus nr 19. In this empirical case I follow the route of bus nr 19 and look through the bus windows at the landscape including the buildings any traveller would see just looking to the left or right and I make stops during the trip. The bus trip enables one to consider how the physical transformation of the landscape supports the technological image of Oulu and the enhancement of ICT. I am also interested in how the material construction of this landscape is received by the local inhabitants.

One stop I make is at the technology park of Oulu. It is the oldest technology park in northern Europe and Technopolis Plc who runs the park is the largest in Europe by the number of technology companies it manages, over 6,000 in total, of which a major part is in Oulu technology park and along the airport road (Technopolis, 2004). In the technology park there is the SMART house in which the future of ICT is discussed and developed. I have taken part in many meetings and discussions in this building. I am especially interested in how internal and external elements are mixed in the technologised landscape made up of the buildings and the activities that happen there.

I end the journey at the University of Oulu, in my office on the university campus section K, third floor, room 319. My particular interest here is in the everyday technology landscape that appears on my working desk and that I feel through all of my senses. I spend a lot of time in the office in face-to-face communication with my personal computer(s). Being surrounded and connected to ICT I participate in human-technology actor-networks.

SCALE AND EVERYDAY LANSDCAPES OF TECHNOLOGY

In geography it has been common to separate place from space and to consider place as the manifestation of the concrete local and space the abstract global in which exists a nested hierarchy of scales (Johnston, 2000, p. 726). Place is the concrete physical environment but space is a somehow absent virtual flow. A landscape is something in-between them. In the past years this clear-cut distinction between local place and global space has been contested (Massey, 2004). More attention has been paid to scale as a connecting concept (Brenner, 2001). Scale is a useful concept to grasp the deworldment processes in which places and spaces are continuously deconstructed and redefined.

In contemporary towns and regions the global is always present in the local and vice versa. Mitchell (2003, 791) has argued that for landscape geography a theory of scale is crucial to understand the histories and geographies of particular places, landscapes and processes working at smaller and larger scales. A more flexible use of scale helps to locate one's position in the world of blurring binaries that are transforming the landscape, for example, human / machine, local / global, and past / future (Gibson-Graham, 2002; Massey, 2004). Scale is a key because much of the current landscape transformation in the era of a new cultural economy is about the politics of scale: which aspects of scale are most attractive to capital and what facilitates the process of 'jumping scales' in which political claims and power established at one geographical scale can be expanded to another (Smith 1993; Johnston, 2004, p. 726). ICT are 'trans-scalar' reaching from bodily presence to imagined human existence in the universe being facilitator, mediator and outcome of our contemporary society of economic globalism.

The economy is a cultural site as indicated by Terkenli (2005). Even so called global capital has a sense of place because (global) capital(ism) is a cultural practice carried out by bodies located in particular places (Massey. 2004, p. 7). Despite the growth of significance of symbolic goods, e.g. flows of virtual economic transactions in the stock exchange, materiality plays a fundamental role in the organisation of the physical landscape. For economic activities to happen at a certain moment something material – location, product, built environment, etc. – is needed. Landscape is in fact fixed in place, relatively stable and solid, of real material force (Mitchell 2001, 278). The temporary placements of things leave their marks on cities and regions (Amin 2004, 34). This is a critical remark towards over-representation of landscapes. Landscape, however, cannot be reduced to the material and visual, as Olsson (2003, 300) underlines: "any understanding of a landscape must consider not only trees and waters, stones and birds, but also hopes and fears, joys and griefs; not only smell and taste but pride and hate; not only light and shadow but power and submission."

Binaries and a dialectical approach (local/global, nature/culture, human/machine, us/them, agency/structure) were topical for geographical research of landscapes in the 1970s and 1980s. The 1990s witnessed a period of criticism towards binaries. Instead there were "trialectics", i.e. third space, in-betweenness, multiplicity of landscapes, etc. (Soja, 1996). Local and global were brought together in glocalisation as a rescaling of the political-economic space, which became an important topic in geography and the social sciences (Swyngedouw, 1992). The early discovery of the "glocal" did not problematise enough the above mentioned separation between the "concrete local place" and an "abstract global space". For example, according to this viewpoint everyday landscapes were concrete local places on which rehierarchised abstract global spatial flows left traces as material manifestations of global capitalism. Currently, the distinction between place and space is becoming blurred. The relational character of space leads to the fusion of local and global. Therefore the earlier understanding of 'glocal' as the conceptual bridge connecting local and global should be criticised because glocal refers to spatially separable local and global. ICT can be used as an empirical example to indicate how such separation should be criticised.

As Doreen Massey (2004, pp. 7-8) points out, space is relational, internally complex, essentially unboundable in any absolute sense and inevitably historically changing. Space is a product of practices, trajectories and interrelations. Space is created through interactions, connections and embodiments. Therefore space is a necessity. Always when we think, talk or write about something we make space – and we do it somewhere. Space is no longer a nested hierarchy moving from global to local (Thrift, 2004). Therefore one could consider scale as a key geographical concept in the early 21st century and a tool to analyse the everyday environment. However, scale has become a 'messy' concept because of the omnipresence of space and the relational understanding of space.

The traditional way to bound space into hierarchic layers overlapping each other is understood today in much more complex ways. Scale is socially produced so it is impossible to categorise scale in fixed terms (Smith, 1993). Scale is processual and dynamically re-scaling (Brenner 2001, p. 592). For Latour (1996), scale is the relational space of actor-networks: fibrous, thread-like, wiry, stringy, ropy, organised as a network of capillaries. The scale of activities has become important in the conceptualisation of relations between places and activities. To study these complex interrelationships one can use flexible scale-jumping that allows the use of scale as an interpretative framework to indicate simultaneously present connections of politico-economic activities and socio-economic struggles in towns and regions (Gibson-Graham, 2002).

The question of scale is no longer a discussion about globalisation and local responses to it but about rapid and flexible scalar changes that have attracted the attention of scholars to understand "what is taking place" (Brenner, 2001). In political activity scale is significant because, according to Harvey (2000), it is necessary to trespass the discursive regimes that separate the local from the global to redefine in a more subtle way the terms and spaces of political struggle in these extra-ordinary times of a new cultural economy.

TECHNOLOGISED LANSCAPES FROM OULU, FINLAND

Cities and regions are agglomerations of heterogeneity locked into a multitude of relational networks of jumping geographical nodes (Amin, 2004). There are increasing similarities in the physical appearance of our cities and regions and in their flexibly scalar landscapes. In the following I illustrate how technologised landscapes have become a taken-for-granted part of my everyday landscape and how I am with many others becoming a cyborg. The technologised landscapes I discuss– the road from the airport of Oulu to the technology park, the SMART building in the technology park of Oulu and the desk with technological devices in my office at the University of Oulu – are my everyday landscapes that exist flexibly and simulaneously at different scales. The bus trip from the airport to my work illustrates how I am surrounded by and taking part in various human-technology actor-networks.

The airport road

The airport of Oulu is located 10 kms southwest of the centre of Oulu in the municipality of Oulunsalo. The airport road runs straight through Oulunsalo (3 km) and Kempele (1.3 km) and then turns north to Oulu. After the centre of Oulu the road continues for a few kilometres to the University of Oulu and to the technology park (Figure 1). This route is also used by bus nr 19, from the very early morning until late at night connecting the two ends of the road.

Logistically, the airport of Oulu is fundamental for the economic activities in Oulu and its surroundings. The next town of over 100,000 inhabitants is located at a distance of 400 kilometres and the national capital Helsinki is 550 kilometres away.

Obviously, the airport is busy: from Oulu there are twenty daily flights to Helsinki, four to Stockholm, two to Tampere, two to Kemi, and occasionally to Russia. Furthermore, seasonally there are weekly charter flights to several southern holiday destinations.

The airport is the second largest in Finland in terms of passengers who number roughly 650,000 annually. The vast majority of passengers travel for business and many due to ICT activities. Some people working for Nokia commute several times per week between the research and development centre in Oulu and the head quarters in the capital region of Helsinki. University and local authority staff travel to Helsinki, for example, for academic and administrative purposes or to catch flights to foreign destinations. The airport is an important site of arrival and departure for many visitors to Oulu. The road from and to the airport is the first and the last image passengers perceive of Oulu. The airport and the airport road connect people with technologies thus becoming active actants in the organisation of the technologised landscapes in Oulu.

Until the 1980s Oulu was a town of pulp industry. In 1982, a technology park was established and the local authorities proclaimed Oulu a technology town. This was received with suspicion and a smile in southern Finland. Nevertheless, during the 1980s the local university focused increasingly on electronics and the development of technology continued. In 1990, the idea of a particular technology-related highway was presented in the development plan for the Oulu region. The regional authorities argued that the airport road to the technology park should be a showcase to illustrate the technological achievements in Oulu and its surroundings. The road should also have a definitive functional purpose as a site for highly competitive technology enterprises (Oulu region 1990, p. 45).

In 1996, to facilitate the local-global technology transformation, the planning board launched an architectural competition regarding seven kilometres of the airportroad. Before the project, the landscape mainly comprised agricultural fields, forest, detached housing, a strip of storage buildings, and the municipal centre of Oulunsalo in a flat semi-urban landscape. The aim of the architectural competition was to find tools to enhance the landscape, logistics and land-use of the airport road and its immediate surroundings. A broader task was to improve the image of Oulu as a highly developed regional centre. The local technological transformations based on global competitiveness in ICT should be manifest in the local landscape. The architectural competition received tens of proposals. They suggested various ways to improve the visual image of the airport road and to enhance the technological image of Oulu and its surroundings. In general, the technologisation of the landscape was highlighted through construction of technology-related buildings. The relation of technology to the surrounding environment and rural landscape was seen as a local particularity (Äikäs 2001, pp. 228-229).

The local and regional authorities started to market the airport road as "the Quality Corridor". The Quality Corridor has become important in land-use planning for Oulu and its surroundings. It is mentioned in the recent master plans of Oulu, Oulunsalo and Kempele, in the collective plan of the Oulu region, and in the regional plan of Northern Ostrobothnia. The Quality Corridor should be impressive, authentic, collective and internationally attractive and constructed in cooperation between the municipalities along the airport road (Pohjois-Pohjanmaa, 2003; Oulu, 2004; Oulun Seutu, 2004). According to Äikäs (2000, pp. 228-229), the landscape of the airport road plays a crucial role in concretising the symbolic imago resources of Oulu and its region. However, it is difficult to find in the documents what "quality" means except that it is related to technology and it is to be located along the airport road. There are no detailed guidelines about how to achieve this "quality" in the physical landscape.

By the mid-1990s, there were few buildings of technology-related enterprises. Since then the collective image of the road has been enhanced by several new buildings of high-tech enterprises, works of art placed along the road, bus stops, etc. The intensive construction took place not only because of the land-use development projects but also due to the expansion of technology-related economic activities in the late 1990s.

Today, travelling on bus nr 19, one notices over one hundred technology-related buildings on both sides of the airport road. The bus trip to the technology park takes 45 minutes during which one can also see agricultural fields, forests and a semi-urban flat landscape. The contrast between rural and high tech land-uses is impressive. Tens of technology enterprises have built their premises along the road and some contain the main offices of the largest regional high tech companies. Some buildings are located close to each other; others are at a distance of kilometres. The design is often typical contemporary international architecture for high tech buildings with glass, steel and wood façades (Figure 1).

In 2004, a large shopping mall is being constructed. According to the local land-use plans, in the future, technologisation of the landscape will continue despite the reduced turnover of ICT enterprises located in the region of Northern Ostrobothniabetween 2000 and 2003 (Jauhiainen et al., 2004, p. 27). The local authorities of Oulunsalo (currently with 9,000 inhabitants) are developing, in the vicinity of the airport, a site for 6,000 workers and the goal is to gain 3,000 inhabitants more in two decades. One high tech enterprise planned to erect a skyscraper that would be visible for tens of kilometres. However, the project has faced controversy in the eyes of the inhabitants of Oulunsalo and Oulu and the implementation of the project has been postponed.

As mentioned earlier, the local and regional authorities use the airport road and its buildings to enhance the technological image. The road is an everyday landscape for inhabitants and visitors of Oulu. To grasp the way this landscape is perceived, understood and used by ordinary people and what attitude they have toward technology buildings and the technological image of Oulunsalo, a survey was conducted among the adult population of Oulunsalo in 2004. The inhabitants defined how much they agreed with statements regarding technology buildings along the airport road. The opinions presented in Table 1 are based on 207 persons representing a 41.4 per cent response rate to the mailed questionnaire.

Even though it has been less than a decade since the first high tech buildings were sited in Oulunsalo along the airport road almost four out of five (78.3 per cent) adult inhabitants agree fully or mostly that high tech buildings characterise the image of Oulunsalo. Local inhabitants have become integrated into this here/t/here technologised landscape without particular guidance. The majority (60.3 per cent) of

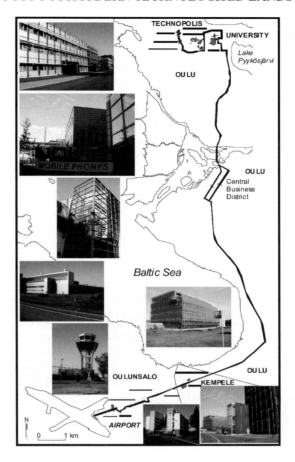

Figure 1. Technologised landscape I. The airport road / the Quality Corridor from the airport of Oulu to the technology park Technopolis and to the University of Oulu. Photos illustrate some technology-related buildings along the bus 19 route, source: author

answers agree fully or mostly that high technology buildings represent a pleasant environment. However, the opinions regarding the success of the location of high technology buildings vary.

It is impossible to separate local from global in this high technology landscape when the scale continuously jumps regarding production and consumption. Regardless of the "inauthenticity" of such high tech buildings globally, three out of four (76.8 per cent) agree fully or mostly that these buildings are impressive. Only one out of fifteen (6.3 per cent) do not agree with the originality of the buildings. Local inhabitants see and feel the presence of the technology-related buildings and are connected to the landscape through the visibility of the buildings.

Table 1. Opinions of Oulunsalo inhabitants on hi-tech buildings (%; n = 207)

statement	I agree fully	I agree mostly	I agree partly	I do not agree	I do not know	total
Hi-tech buildings characterise the image of Oulunsalo	28.5	49.8	17.4	1.0	3.4	100
Image of Oulunsalo leans too much on technology	7.7	24.2	40.1	14.5	13.5	100
Location of hi-tech buildings is successful	7.7	24.2	40.1	14.5	13.5	100
Hi-tech buildings are original	17.4	41.1	29.0	6.3	6.3	100
Hi-tech buildings are impressive	27.5	49.3	15.5	4.8	2.9	100
Hi-tech buildings represent a pleasant environment	15.9	44.4	25.1	12.1	2.9	100
Hi-tech buildings fit into history and tradition	7.2	20.8	33.3	18.8	19.8	100

It is the past that makes the local inhabitants a bit more sceptical. Two out of five (38.6 per cent) do not know or do not agree that high tech buildings fit into history and tradition. Also one third (31.9 per cent) agrees fully or mostly that the image of Oulunsalo leans too much on technology. People are aware of high tech buildings even though only a few local inhabitants ever enter these buildings but many consume the applications produced in these buildings. This is one way also that local inhabitants become involved in human-technology actor-networks.

The SMART house in Technopolis and the business breakfast

The case of the airport road exemplifies how the external reconstruction of a technologised landscape took place and how local inhabitants were integrated into it. To illustrate briefly the internal and closer dimension I continue with bus nr 19 to Technopolis, the technology park of Oulu. One may enter freely the technology park but it is not obligatory for ordinary visitors of Oulu. Technopolis is located five kilometres north from the town centre between the north-south motorway and the University of Oulu. The site is part of the Quality Corridor and its visual image is important. What is done inside the buildings, however, is more significant. It is commonly argued that the technology park and the university have been crucial for the technological transformation of Oulu.

The technology park was established in 1982 and it has been frequently expanded. The private stock exchange rated enterprise Technopolis Plc is the main operator in the technology park. The purpose of the company's operations is to promote the business of high tech companies by utilising high tech competence developed through cooperation between companies, universities, polytechnics and research institutes and by providing operating environments for the implementation of product and business ideas in the Oulu region and in the Helsinki metropolitan area. The business idea of Technopolis is to control real estate on the basis of ownership and leasing rights and to construct operating and service premises that are then rented out to companies. It also provides rental equipment, training and advisory services in the high tech area as well as project and service operations promoting the business of customer companies (Technopolis 2004).

Today Technopolis occupies a territory of 60 hectares with building permission for 150,000 cubic metres in the technology park of Oulu: currently there are 29 buildings in which 1.7% of the working premises are vacant. In the park there are 172 high tech enterprises with about 4,500 employed and a further 1,500 along the airport road. The enterprises work in communication, electronics, micro-, nano-, bio, welfare and environmental technologies, especially in mobile applications (Technopolis 2004). Technopolis is a privately-run company but the City of Oulu is its second largest owner with 18.1 percent of shares. Furthermore, the local authorities are involved in the technological park through the maintenance of the road network, the sewage system, etc.

According to the managers of park activities, the principal street that runs through the technology park, called Elektroniikkatie (Street of Electronics), is globally most important regarding "Symbian", the leading technology for mobile communicators (Launonen, 2004). One stop before the terminal of the bus nr 19 route, at Elektroniikkatie 8, there is a building called the SMART house (Figure 2). The SMART house is a a mere part of the everyday landscape of people in Oulu, but it is very important for those who work in the technology park. The building has 7,800 square meters of floor space, hosting 18 enterprises with 250 staff of which 210 in technology-related activities. Besides the normal day-to-day work of technology enterprises, business meetings, seminars, and gatherings for technology development are organised there.

The SMART house is owned by Technopolis. It was designed by the Finnish architect Pekka Laatio / Laatio Architects and constructed in 1999 – 2000. "Smart" has a double meaning: the building itself has certain advanced technology but it also provides a meeting place for people involved in technology development. During construction particular attention was paid on large internal open spaces (Figure 2). The managers claim that one of the most important sites for technology development in the whole technology park is the lobby and the restaurant of the SMART house. One meets there colleagues, business partners, and sometimes strangers and one talks about technology-related topics (Launonen, 2004). As Michael Porter (1996) claims, positive chance is one important factor in the creation of more competitive enterprises, regions and nations. Therefore the internal space in the SMART building is currently used with this purpose in mind.

One does not need specific permission to enter the SMART house unlike almost any other building in the technology park. However, local inhabitants seldom enter the building: daily visitors are technology developers, administrators and businesspersons. One way to experience scale-jumping in the SMART house is to visit the "business breakfasts" that have been organised almost monthly for the past years. The concept has been taken from certain US technology parks in which business breakfasts have proved to be important and thought-provoking (formal and informal) meetings between technology-related businesspersons. While having a real breakfast at large tables one can share business information, listen to lectures and enjoy the opportunity to meet and discuss about technology-related issues (Saukkonen, 2004).

In the SMART building one has to register to attend the business breakfasts but there is no registration fee. About 100–150 participants normally take part in the events. One enters the technology park early in the morning and proceeds to the SMART building and the main open space of the building. One takes food from a buffet table and sits down at one of many tables. There are formal short business briefs and some discussion among the participants. Meanwhile some people send text-messages and quietly take in calls. The discussion among the participants varies: it might be about the quality of the breakfast, finance problems of local biotechnology enterprises, international law on copyright or future challenges in the mobile ICT. People come together from different places, meet among colleagues and strangers at the breakfast table and create a social space. An ordinary breakfast is transformed into a physical and discursive technologised landscape with multiple scales. However, most of the participants are quiet, hardly anyone asks questions and comments on the presentations are few. Well, it might be difficult to lobby at 8 am – either everything is self-evident and expressed in silence or it is just too early. In my opinion, people fuse in this discursively created, materially constructed and scalarly omnipresent technologised landscape, which they are not able to critically question.

The office desk with (and its?) technological devices

The third case narrows the technologised landscape to the micro-scale, namely to my office and to the technological devices on my working desk. My office is reachable by bus nr 19 and by a few hundred meters walk inside the university campus. The campus at the University of Oulu is a network of buildings with a distance of over 1.5 kilometres from one corner to another materially and metaphorically under the same roof – a particular actor-network itself.

Despite this mobile ICT I still spend most of my working time in one place, in my office at the University of Oulu. The office for writing is 3 x 4.6 metres. Two walls are covered by full book-shelves, one wall has space for clothes and one wall has windows facing a forest outside. A U-shaped desk sits in the middle of the room. On the table towards the door there are papers that require immediate attention and on the wall-facing side, piles of paper that I should have paid attention to. The side facing the window has a scanner, a printer, a personal laptop and in the corner a personal computer (PC). My office is not yet a wireless one so there are 13 cables connecting the devices to the rest of the world. This is my everyday landscape of technology: surrounded by and connected to many machines during the day.

Figure 2. Technologised landscape II. The SMART house, source: author

Normally, from Monday till Friday, I arrive at my office around 9:30 am and I leave it around 6 pm – sometimes later. The very first thing is to switch on the PC, then to place the lap top on the desk and to switch it on. Then I browse through the e-mails and answer those that are urgent. I have two sites to look at, one for the University of Oulu and another for the University of Tartu where I work part-time. After that I normally continue writing some papers, preparing lectures with Microsoft™ PowerPoint™ and browsing some information from the Internet with Google™. On the screen of my PC there are icons of programmes and in the background a green field (Figure 3). This technologised landscape never changes regardless of the seasons in Oulu – until the ICT devices will be replaced by more advanced ones. However, the new technologised landscape will probably have the same functions, the same landscape on the screen, and the same icons.

Computers are an inseparable part of (academic) life nowadays and mobile ICT devices are becoming inseparable from our body. The ICT shapes our everyday contacts: I have more frequent contacts with people in spatially distant places than with colleagues in my immediate physical environment at the department. A characteristic of current life is that white-collar employees are more often in active "face-to-face" contact with the PC than with their spouse, children, not to mention relatives, friends or colleagues. I am in immediate face-to-face connection with my lap top and PC normally about five-six hours per working day – interrupted only by phone calls, meetings, lunch breaks, and lectures. In this everyday technologised

Figure 3. Technologised landscape III. The office desk with technological devices,
source: author

landscape we are connected to networks through ICT devices but they also serve us: they wake us up, remind us of things to do and expose our daily spatio-temporal patterns for location-based services and surveillance. Increasingly, people can measure in just hours the moments when they have not had access to information and communication networks. People are melting into the technologised landscapes.

CONCLUSIONS

In this chapter I have not proposed any general theory for technologised landscapes. Nor have I examined the landscape through concepts of capital circulation, race, gender or power or specified how the cultural, social, political and economic are folded into the landscape. Instead, the case study of Oulu is a diagnosis of contemporary technologised landscapes in which we live in the form of three moments. From my point of view, *Zeitdiagnose* is one real possibility to overcome the confrontation with morphologists and representationalists in landscape studies. Courageously openly used it enables us to understand the folding of technologisation into the landscape. We can also grasp how any activity is a translation from one specific action and experience to another. Technologisation brings people and things together but it is up to the practicioner of *Zeitdiagnose* to unravel this hybrid relationship. Therefore the richness of *Zeitdiagnose* is in its "giving a supportive hug" to continuously changing geodiversities in the processes of deworldment.

Scale is flexibly present in the technologised landscapes of Oulu as assemblages of more or less distanced economic relations with different intensities at different locations (Amin & Thrift, 2002). In the technologised landscapes studied, the airport road, the SMART house, and the ICT items on the office desk, do not exemplify nested hierarchies from global to local but a multitude of relational networks. The difference between here and there becomes unclear. Like space (Massey 2004) technologised landscapes are relational, products of practices, trajectories and interrelations created through interactions, connections and embodiments.

Physically seemingly similar technologised landscapes (icons on the PC screen, ICT devices, physical technology networks, component producing buildings, R&D laboratories, imago promoting brochures, etc.) are "inauthentic" in their visual appearance globally but one should carefully contextualise these landscapes. Somehow technologised landscapes are "actors" that are not substitutable for anyone / anything else, unique events, totally irreducible to any other (Latour, 2004). Place-space relations are continuously deconstructed and redefined in the process of deworldment.

The physical landscape plays an important and active part in changing the relationship between people, technology and environment. The physical landscape along the airport road significantly promotes the myth and image of Oulu's "technological miracle". Filling the landscape with technology buildings makes the landscape a site of affirmation of the dominant narrative of the technological supremacy of Oulu and of the ICT cult without any resubjectivisation of the narration. The myth of the "miracle" is visualised in the printed media in Finland and on the Internet through the representation of the material technology-related landscape of Oulu. This further affirms the dominant narrative of Finland as the leading society of the ICT revolution. The business breakfasts in the SMART house are more self-legitimising consensual gatherings to promote the ICT cult than to critically self-reflect on its challenges. The result is that on my desk there will be more and more ICT devices to get acquainted more intensively with the new (global?) cultural economy.

In post-postmodern landscapes scales of technology and mind jump flexibly. It is difficult to separate oneself from technologised landscapes. Technologised landscapes are always t/here surrounding, intertwining and penetrating us. The binaries between human / machine, local / global, past / future are blurred in the unified sp/pl/ace of technologised landscapes. A characteristic for contemporary politics of technologised landscapes is that the authenticity of narratives does not need to be shared by many people. The local inhabitants are lured into the technologised landscape by copying and pasting technological traces in the "local" landscape regardless of problems in their "authenticity". People get used to this landscape and capital for technology development can be more convincingly lured when there is something material and visual in the landscape. The landscape along the airport road is made to work to claim cultural authority; it generates profit, and it reconstructs a particular social identity. Exchange and use values melt into one commodity through subtle modes of attraction, inclusion and exclusion.

The *Zeitdiagnose* of Oulu can be used to grasp, understand and envision the continuously passing moments of technologised landscapes. The case is a small subjective one of history and geography regarding particular places and processes of ICT simultaneously present at all scales. It is a selective narrative based on varied empirical material and observations. The participatory observation has helped me to find my place and time in this story. The airport road in Oulu in its complexity is a technology actor-network, a particular hybrid unsubstitutable human-machine actant, and I have become a cyborg.

I admit that the relationship between conceptual construction, empirical evidence and conclusions is complicated but I believe in the power of subjective synthesis. This conclusion should not be understood as a fierce critique against the growing presence and dominance of ICT in our life. However, it is an openly committed and worried viewpoint and statement about this "post-postmodern" period when technologised landscapes are inscribed in our bodies more tightly than a tattoo. Faster and more comprehensively than automobiles in the early 20th century and television in the mid-20th century, the ICT devices are being "naturalised" in our life right now. Technologised landscapes transform our time-space practices but the need for un-technologised or de-technologised landscapes remains unanswered in this paper. What is then the normative political statement of *Zeitdiagnose* concerning the pathologies of technologised landscapes? I do not know: astonished I step aside, disconnect the ICT, and switch myself off...

Department of Geography, University of Oulu, Finland & Institute of Geography, University of Tartu, Estonia.

ACKNOWLEDGMENTS

This research has been funded by the Academy of Finland Research Grant 200771

REFERENCES

Äikäs, T. (2001). Imagosta maisemaan. Esimerkkeinä Turun ja Oulun kaupunki-imagojen rakentaminen. PhD dissertation. *Nordia Geographical Publications*, 30 (2).

Amin, A. (2004). Regions unbound: towards a new politics of place. *Geografiska Annaler B: Human Geography*, 86 (1), pp. 33-44.

Amin, A. & Thrift, N. (2002). Cities reimagining the urban. Cambridge: Polity.

Backhaus, G & Murungi, J. (2002). Introduction: Landings. In G. Backhaus & J. Murungi (Eds.) *Transformations of urban and suburban landscapes. Perspectives from philosophy, geography, and architecture* (pp. 1-18). Oxford: Lexington Books.

Brenner, N. (2001). The limits to scale? Methodological reflections on scalar structuration." *Progress in Human Geography*, 15 (4), pp. 525-548.

Gibson-Graham, J.-K. (2002). Beyond global vs. local: economic politics outside the binary frame'. In A. Herod & M. Wright (Eds.) *Geographies of power. Placing scale* (pp. 25-60). Oxford: Blackwell.

Haraway, D. (1991). *Simians, cyborgs and women – the reinvention of nature*. London: Free Association Books.

Harvey, D. (2000). *Spaces of Hope*. Oxford: Blackwell.

Henderson, G. (2003). What (else) we talk about when we talk about landscape: for a return to the social imagination. In C. Wilson & P. Groth (Eds.) *Everyday America: cultural landscape studies after J.B. Jackson* (pp. 178-198). Berkeley: University of California Press.

ITU (2004). International Telecommunication Union statistics. <http://www.itu.int/ITU-D/ict/statistics/> 20 August 2004.

Jameson, Frederick (1991). *Postmodernism – or the cultural logic of late capitalism*. London: Verso.

Jauhiainen, J. (2003). Learning from Tartu – towards post-postmodern landscapes. In H. Palang & G. Fry (Eds.) *Landscape interfaces. Cultural heritage in changing landscapes* (pp. 395-406). Dordrecht: Kluwer.

Jauhiainen, J., Ala-Rämi, K. & Suorsa, K. (2004). *Multipolis – teknologian, osaamisen ja kehittämisen yhteistyöverkosto*. Helsinki: Sisäasiainministeriö Alueellinen kehittäminen.

Johnston, R. (2000). Scale. In R. Johnston, D. Gregory, G. Pratt & M. Watts (Eds.) *The dictionary of human geography* (pp. 724-727). Oxford: Blackwell.

Latour, B. (1993). *We have never been modern*. Hassocks: Harvester Wheatsheaf.

Latour, B. (1996). *Aramis: or the love of technology*. Cambridge, MA: Harvard University Press.

Latour, B. (1999). On recalling ANT. In J. Law & J. Hassard (Eds.) *Actor network theory and after* (pp. 15-25). Oxford: Blackwell.

Latour, B. (2004). A Dialog on Actor-Network-Theory With a (Somewhat) Socratic Professor. In C. Avgerou, C. Ciborra & F. Land (Eds.) *The Social Study of ICT*. Oxford: Oxford University Press.

Launonen, M. (2004). Director of the Programme for Oulu Region Centre of Expertise in Technopolis. Interview, 25 May 2004.

Massey, D. (2004). Geographies of responsibility. *Geografiska Annaler B: Human Geography*, 86 (1), pp. 5-18.

Matless, D. (1998). *Landscape and Englishness*. London: Reaction Books.

Mitchell, D. (2001). The lure of the local: landscape studies at the end of a troubled century. *Progress in Human Geography*, 25 (2), pp. 269-281.

Mitchell, D. (2003). Cultural landscapes: just landscapes or landscapes of justice? *Progress in Human Geography*, 27 (6), pp. 787-796.

Noro, A. (2004). Aikalaisdiagnoosi: sosiologisen teorian kolmas lajityyppi? In K. Rahkonen (Ed.) *Sosiologisia nykykeskusteluja* (pp. 19-39). Helsinki: Gaudeamus.

Olsson, G. (2003). Landscape of landscapes. In J. Öhman & K. Simonsen (Eds.) *Voices from the north: New trends in Nordic human geography* (pp. 299-308). Aldershot: Ashgate.

Oulu region (1990). *Regional plan for Oulu region* [in Finnish]. Oulu.

Oulu (2004). *Oulun yleiskaava 2020* [master plan of Oulu for 2020 in Finnish]. Oulu: Oulun kaupunki.

Oulun seutu (2004). *Oulun seudun kuntien yhteinen yleiskaava 2020* [master plan for the urban region of Oulu 2020 in Finnish]. Oulu.

Pohjois-Pohjanmaa 2003. *Pohjois-Pohjanmaan maakuntakaava*[regiona l plan for Northern Ostrobothnia in Finnish]. Oulu: Pohjois-Pohjanmaan Liitto.

Porter, M. (1996). *The competitive advantage of nations*. London.

Saukkonen, J. (2004). Coordinator of Global Software Forum in Technopolis. Personal communication., 8 June 2004

Smith, N. (1993). Homeless/global: Scaling places. In J. Bird, B. Curtis, T. Putnam, G. Robertson & T. Tickner (Eds.) *Mapping the futures: Local cultures, global change* (pp. 87-119). London: Routledge.

Soja, E. (1996). *ThirdSpace: Journey to Los Angeles and other real-and-imaged places*. London: Routledge.

Swyngedouw, E. (1992). The mammon quest. 'Glocalization', interspatial competition and the monetary order: The construction of new scales. In M. Dunford & G. Kafkalis (Eds.) *Cities and regions in the new Europe. The global-local interplay and spatial development strategies* (pp. 39-67). London: Belhaven Press.

Technopolis (2004). Technopolis articles of association. <http://www.technopolis.fi/index.php?area_id=4&lang_id=1> 20 August 2004.

Terkenli, T. (2005). Landscapes of a new cultural economy of space: an introduction. In A-M d'Hauteserre and T. Terkenli (Eds.) *Landscapes of a new cultural economy of space*. Dordrecht, Kluwer.

Thrift, N. (2004). Intensities of feeling: towards a spatial politics of affect. *Geografiska Annaler B: Human Geography*, 86 (1), pp. 57-78.

MARTINE GERONIMI

SYMBOLIC LANDSCAPES OF VIEUX-QUEBEC

World heritage plaque in front of Port Museum, 2005, Old Québec, source Martine Geronimi

"As long as Quebec City, through the determination of its sons and daughters, expresses in its invincible individuality the unity and community of the human adventure, it can be proud of contributing to the rapprochement of all peoples on Earth[1]"
1985 UNESCO Deputy Director-General, cited by Historian Gaston Deschênes in Forces N°117, 19973kmj21111ikj

INTRODUCTION

Sociologist Sharon Zukin points out the emergence of a symbolic economy in her book The Cultures of Cities in 1995. She demonstrates how in today's capitalistic society the global economy re-appropriates culture. It seems that in our fast

T.S. Terkenli & A-M. d'Hauteserre (eds.), Landscapes of a New Cultural Economy of Space, 213-237.

changing world, culture produces symbols for the global market. In this chapter I would like to demonstrate how World Heritage landscapes reveal this new symbolic economy. As a cultural geographer, I consider that we are facing a new approach of space not only centered on the uneven geography of costs and revenues (Thrift, 1986; Porter and Sheppard, 1998) but on a culture-centered approach (Terkenli, 2002; Sack, 1997). In this way, Heritage and Tourism become cultural industries revealing the problems of our post-modern society that commodifies cultural heritage as a tasteless product. The Past and Architecture have now been transformed into a simple marketed tourist item with a touch of nostalgia and even irony.

Vieux-Québec in Quebec City is a perfect example of the collapse of barriers in space and time and a progressive loss of distinctiveness. The proclamation of the French identity of the old district is reappropriated by tourism, by its residents and by business interests.

We will develop at the end of this chapter the case of reappropriation of typical francophone symbols. In this instance, we will present anglophone businesses in Guelph (in the province of Ontario) who have transformed Old-Quebec into a commercial brand.

The attribution of World Heritage label in 1985 by Unesco to the Vieux-Québec has led to the integration of a new vision of the city inhabitants. Nevertheless, the old district seems to live artificially, as it becomes a symbolic landscape and a subject of the spare-time economy. I intend to illustrate this point of view by comparing the results of surveys sent out in 1990 and 2000, questioning Vieux-Québec residents on their relation to the heritage and tourism of the area. The second point will come from an analysis of the city's marketing discourse on the Web. I am interested in identifying the processes of transworldment by examining how French imageries of Vieux-Québec are disseminating around the world electronically and reappropriated by residents themselves and by investors.

SYMBOLIC ECONOMY

In our 21rst Century, three points, which are the urban sphere, the production of symbols as basic commodities, and the production of spaces, exemplify the symbolic economy. In the public urban sphere, this symbolic economy refers to the process of wealth creation through cultural activities, including the fine and performing arts, creative design and sport. Culture plays an increasingly important role in the design and planning of cities. The new economy produces sites and symbols of the city and of culture.

Sharon Zukin defines symbolic economy as "the intertwining of cultural symbols and entrepreneurial capital" (Zukin, 1995, p. 3). The result of the decline and relocation of manufacturing facilities left a gap in many traditional industrial economies. This new economy places design and innovation at the forefront using the increase of knowledge-based industries and activities.

The rise of this symbolic economy is associated with recent changes in our societies all around the world (Terkenli, 2002; Meethan, 200; Bourdieu, 1991). We have entered the 'post modern' society in which we give privilege to innovation, mobility, and knowledge. Meethan conceptualizes tourism as "a global process of commodification and consumption involving flows of people, capital, images, and culture". (Meethan, 200, p. 4).

An overflow of messages and images disseminated on the Internet relate to the concept of "symbolic economy". As Castells announced:

> "In a broader historical perspective the network society represents a qualitative change in the human experience. ...Because of the convergence of historical evolution and technological change we have entered a purely cultural pattern of social interaction and social organisation. This is why information is the key ingredient of our social organisation and why flows of messages and images between networks constitute the basic thread of our social structure." (Castells 1996:, p. 477).

During these last twenty years we have witnessed the rise of consumer product industries and multi-media industries that created our global market place. The rise of flexibility in organizations, productions, and specializations involved these changes (Lee & Willis, 1997). These authors show that globalization is a process, which moves us towards a world of mobile and fancy-free capital. In this context, transnational corporations are currently all-powerful agents of change. The financial market forces are dominant, facing insignificant nation-states, homogenizing social, political, and economic conditions of these nations and proclaiming: "globalization is natural". By this, the authors assert that the nation-state is the main political instrument for promoting the ideology of globalization (Dicken et al., 1997, p. 159). However, the nation-state is not alone in this endeavour; other promoters, such as academics and technocrats who produce the theories of this discourse are very business-oriented and provide freely their advice to governments and the large business class (Leyshon, 1997, p. 143). Finally the corporate sector produces its own discourse about globalization. Leyshon asserts that these groups are only interested in developing a corporate strategy in which the effects upon the majority are not taken into account.

In presenting a partial representation of the world, the C.E.Os are able to create a self-affirming discourse. This discourse is simple and gives, at the same time, an explanation, and a program of action to face the complex issues surrounding the inevitability of globalization. Other authors demonstrate that some forms of government intervention are viewed as ineffective and governments themselves are perceived as being "hollowed out." (Dicken et al., 1997, p.159). These narratives legitimate neo-liberal policies and introduce market principles into every area of economic life and, thus, affect every citizen.

According to Lury (1996), culture itself is significant in the formation of social and political groups. Culture is instrumental in the defining and forming of an individual's self-identity. The author draws attention to the branding and the marketing of places. In our capitalistic world dominated by mass commodification, producers attempt to distinguish their goods by infusing them with social meaning. A major problem in our commodified world is authenticity and "authenticity" is a particularly desirable disguise for mass produced goods. In heritage places, tourists are searching for this same authenticity and Lury opens our eyes about the problems of staged authenticity. As long as tourists will just want a sight of the local atmosphere, a glimpse at local life, without any knowledge or even interest about the real life of the place, staging will be inevitable.

Allen Scott (1997, p. 323) points out that

> Capitalism itself is moving into a phase in which the cultural forms and meanings of its outputs become critical if not dominating elements of productive strategy, and in which the realm of human culture as a whole is increasingly subject to commodification, i.e. supplied through profit-making institutions in decentralized markets. In other words, an ever-widening range of economic activity is concerned with producing and marketing goods and services that are infused in one way or another with broadly aesthetic or semiotic attributes

To avoid this aesthetic or semiotic cover-up, cities need to individualize themselves and become unique in order to put their name on the map. In order to do so, they need to create symbols and market themselves as a symbolic product. In western countries government policies gradually came to focus on the support of culture of cities. The governmental support for culture is seen as a way of enhancing the reputation of these specific places in order to revive their economic prospects (Meethan, 2001; Florida, 2002). It is sold to the citizens as the best means to improve their quality of life thanks to the revenues generated and the growing renown of the country.

We are consciously, or not, participating in a revolution. By this, process we are reversing the role and meaning of culture. Today we consider culture as a creator of wealth, which was not the case in the past. Before culture was a consequence of wealth. This revolution is supported by images. We live in a mediated world. The representations of cities, their images influence more and more consumer decisions (Morgan & Pritchard, 1998 and 2001). Virtual landscapes are produced by the Internet and they create an imaginary city for tourists and migrants.

In Canada, the French part is recognized officially as the Province of Quebec. It owns two virtually marketed cities: the festive Montreal and the French historic city of Quebec. Montreal is the second Canadian economic metropolis after Toronto. Quebec is the political center considered by French Canadians as the future capital of a future independent nation, free Quebec, and by the English Canadians as the Old Capital. On the other hand, Canadian and Quebec governments both participate in the branding of the two cities. They both understand that, in our time of increasing mobility, tourists are very important for the regional and national economy. However they are a rather volatile clientele that must constantly be renewed and attracted by symbolic images (Geronimi, 2003)

The symbolic economy is the embodiment of culture and past which have been reappropriated by the business world. Heritage Tourism can be defined as a complex activity that links a recomposed past, international mobility and political and economical interests situated in the cities' old center (Geronimi, 2003). Baudrillard (1981) wrote that capitalism is now concerned with the production of signs, images, and sign systems rather than commodities themselves. In our new 21st century, images are constantly present and Vieux-Québec is a model of this Cultural Revolution we participate in: the *Revolution Tranquille* (The Quiet Revolution[2])

(Rocher, 1973; Bouchard, 1996; Courville, 2000). By appropriating the *Revolution Tranquille* into Quebec City, we consecrate its revered political and historical identity for the inhabitants (the Quebeckers). Moreover, Old-Quebec has become to international tourists a high-end commodity[3.]

VIEUX-QUEBEC AND UNESCO WORLD HERITAGE LABEL

If the production of space depends on symbolic considerations as expressed by Zukin (1995), pressing questions for these historic places are: what and who should or should not be communicable? In this case, the French past is the city's major attribute. Three characteristics create a distinctive city: the geographical site, the built environment and the people appropriating this specific space. In Quebec, the site is famous and spectacular with the magnificent St Lawrence River dominated by the Cap Diamant (Cape Diamond). The built environment is renowned for its fortified walls and the Frontenac Hotel. Together with its French-speaking inhabitants they represent the city's uniqueness. The old part of Quebec City is divided in two sections: Lower Town and Upper Town. The Lower Town commonly named Place Royale is very interesting. Its main attraction is the restored and reshaped image of the cradle of French civilisation in North America or cradle of Francophone culture in North America.

In December 1985, Quebec received the label of Unesco World Heritage. It was the beginning of international recognition of Quebec both as a historic city, the only remaining fortified city north of Mexico and as a Francophone destination. Unesco World heritage list is an excellent means to bring attention to the city. The label then becomes a trademark for the city. As soon as a city is recognized as a World Heritage jewel, people flock there because registration on the list guarantees authenticity. Vieux-Québec gained more tourists after the attribution of this trademark as illustrated by the flows of Japanese tourists: before 1985 there were no Japanese tourists and after the attribution of the label, Japanese tourists flocked to Quebec City (Geronimi, 2001, 2003).

Quebec City is a socially constructed place, which carries many symbolic values. The perception of Quebec City by inhabitants and visitors after 1985 has changed. The attribution of an international recognition and the transformation of physical environments as of Place Royale contributed to changes in the city's representations. People who experience the city become subject to the cultural context of Quebeckers' celebration of their identity as a distinct society and by the same means participate in the new perceptions of Vieux-Québec. The visitors then become active participants in this new constructed image or new perception. In our era of world spectacularization (Debord, 1967), we have to be sensitive to strategies producing and promoting these city images. The context of Vieux-Québec demonstrates the city's capacity to promote itself as the capital of a future nation and as an international tourist destination, as both a legitimate political center and an economic success.

Self-advertising is very important for the city. The Internet is an excellent tool to propagate advertising messages worldwide. Quebec City thru the Internet has strengthened its reputation as a successful place. What is very important is that the success story message is not just good enough to attract visitors but can actually attract immigrants and investors. This symbolic imaging is present through the city landscape. The built environment, architecture, and monuments shape the symbolic landscape. For Zukin (1995) architecture is the capital of symbolism. In Vieux-Québec the notions of city improvement and architectural rehabilitation are pre-eminent. Municipal, Quebec and Canadian Governments played active roles in the revamping of the city's image during the second part of the XXth century, especially from 1970 to 2000. Architectural studies that flourished in those years helped to define a frame for changes in order to give choices. Those choices contributed to determining values for the conservation or the destruction of places. French urban values were selected during the *Révolution Tranquille* Era at the beginning of 1960. Modernizing Quebec society, while preserving its French Past, was the most important mission of the Quebec Government. Thus, being selected by Unesco, as jewel of humanity, emphasized Vieux-Québec's sacralization.

Figure 1. Frontenac Hotel, cityscape Lower Town and Upper Town, Martine Geronimi 2003

World heritage landscapes as those of Vieux-Québec are cultural landscapes in which symbols and images of the past are recurrent. This French past celebrates its distinctiveness and uniqueness. Quebeckers are different from Canadian and North American Anglophones; they sustain the French language in North America. This political agenda is very important for the relationship between Quebec Government and their voters. It is also important on the international scene, especially in the Francophonie.

World heritage landscape is a trademark which stimulates cultural industries: Museums (Musée de la Civilisation, Musée de l'Amérique Française, Centre d'exposition de la Place royale) and events, such as Festivals (Fêtes de la Nouvelle-France for instance). They exemplify how cultural industries are directly related to tourism industries. Culture is a product consumed by visitors. In Vieux-Québec, commodification of culture is embedded in the symbolic landscape presented to mass tourists of our globalization era. Millions of tourists flock to this vernacular provincial place attracted by the label 'World Heritage city' and its commoditization and sale on the Internet and in the media. They come to Quebec City for the existing aura of celebrity, an expression forged by Walter Benjamin (1928).

Figure 2. French style decorated Place Royale, Martine Geronimi 2003

The Word Heritage label increases the celebrity of an already well known historical place and validates the authenticity of this place. Therefore Vieux-Québec presents a double interest for tourists; they may first of all discover a famous heritage place, which is authenticated as "historically authentic"; they can then consume the authenticity signs of Vieux-Québec. For example, in Quebec a Brittany creperie will seem typically French for an American tourist, which is entirely false for French tourists. Being that Brittany is a region of France, the common French tourist will easily be able to recognize the copy from the original creperie. We must clarify that indeed French tourists and Francophone residents come from two

Figure 3. Tourists in Treasures Street Upper Town, M. Delisle 2003

entirely different backgrounds. The experience of a quaintly-decorated folk restaurant emphasizes the eating experience. Tourists test the commodification of urban and cultural identity. It is a sort of folk tale that tourists and residents participate in, more often than not, unconsciously. People, residents or tourists have been convinced of the authenticity of their experience by the preserved or rebuilt environment and the mediated representation of Vieux Québec (Geronimi, 2003; Dann, 2002).

VIEUX-QUEBEC RESIDENTS SURVEYS

Vieux-Québec, like many restored historical cities, represents old centers that are becoming depopulated only to have the cities activities later reappropriated for tourism and a leisure consumption economy based on urban decoration and entertainment. With the expansion of the first francophone University to the new suburbia of Sainte-Foy during the 1950s, Vieux-Quebec was left desolate. University of Laval, created within the walls of the Grand Seminaire in 1852, has now become too small for the ever-increasing number of students. Its only solution was to follow the new North American trend of moving to suburbia and taking its faculties and colleges along to Sainte-Foy, leaving a gaping economic hole in the Vieux Québec. During the 70s, the Vieux Québec was renovated based on its French historical heritage to accommodate the growing activity of tourism.

Since 1985, when the world Heritage city label conferred additional prestige on the Vieux-Québec, the city has seen an increase in the number of its international

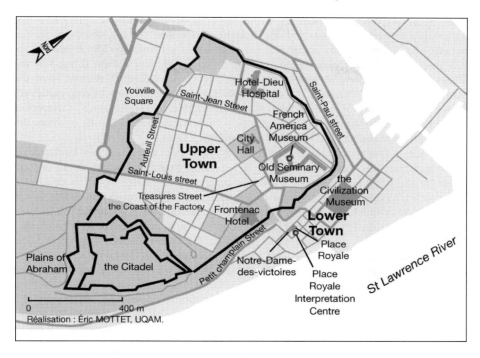

Figure 4. Map of Vieux-Quebec-Eric Mottet 2004

customers. Beyond the superficial and temporary appropriation of Vieux-Québec by the tourists, we need to remember that residents occupy this place on a more permanent basis.

Consumption by contemplation and ritualisation combines with the Unesco recognition to make the city a sort of sacred place. The Vieux-Québec Citizens Committee initiated two great investigations during the decade 1990-2000. The present chapter depends on the results of the two surveys that allow for a lively interpretation of the question of relations between heritage, residents, and tourism[4].

These surveys help us to answer to the main questions that are:

Who resides in the old center of Quebec? What do residents seek in a heritage place? Were their expectations met or not? How do they perceive their contact with the tourists? What then is the impact of the interaction between tourists and a community of strong heritage on its social and heritage landscape? Did their expectations change with time? What is the image of a heritage city?

The cartography of Vieux-Québec highlights the geographical properties of the site which divides it into two parts. It shows a significant physical demarcation in the landscape between the Upper-Town and the Lower-Town, which we take into account to determine the city's sociological profile. Maranda (1991) adhered to the idea that demographic factors corroborated this topographic and physical distinction (see figure 4). Nevertheless, he wrote that social life and citizens erase the dichotomy between the top and the bottom of Quebec's old Center-city to the benefit of a homogeneous community, a true fellow-citizenship (Maranda, 199, p. 23). In 2002, Simard, a town planner, insisted that the distinct society formed by the Vieux-Québec residents is an act that goes beyond the "distinct society" that the other inhabitants of Quebec City as a whole pride themselves of having achieved. However, this study also underlined the demographic and sociological differences associated with the top and bottom physical dichotomy. These results confirm the conclusions reached using GIS in 2001 (Geronimi, 2001).

The Vieux-Québec is now repopulating. The prospect of a fossilized landscape is no longer true. The downtown sector is regaining residents, thanks to a constant effort by the Mayor since 1994. However uptown is suffering from the continued decrease of its population. In spite of the addition of 439 residential units in the whole of Vieux-Québec, global population increased only by 117 from 5180 in 1991 to 5297 in 1996. In 1961, Vieux-Québec counted 10252 inhabitants which included the 6500 residents in uptown. Forty years later, the Vieux-Québec uptown had lost more than half of its residents with an official number of 2945 residents. Hence population increase in Vieux-Québec has occurred only in its lower part, stimulated by a vigorous housing policy. Since 1994, more than 248 residential units have been added around the port and the Museum of Civilization. The new units represent 24.5 % of the housing market in the sector. 631 people resided downtown in 1981. The city has since gained 1422 residents. As justly pointed out by Lamarche: "What is the critical mass of residents necessary to animate an urban environment" (1998, p. 5)? The cartographic analysis based on GIS offers a reliable, general panorama: The Vieux-Québec is a 'non-place' for families (Augé, 1995). There, married couples are very few and people more than 65 years old are rare. One deals with a population of young and unmarried tenants, sometimes couples, or single parents.

In Quebec city, 51.8% of the population are female, against 48.2% male. The average for males in Vieux-Québec District is 52.5% against 47.5% for females. This male dominance is particularly clear in Uptown and in Cap-Blanc, a portion of Downtown. However, Cap-Blanc lodges a great proportion of families made up of married couples with children.

A more advanced analysis distinguishing owner residents and tenants, demonstrates that families who own their homes reside in downtown Vieux-Québec, while the owners and tenants mix is more heterogeneous uptown. The average value of homes there ranges between Can$71,000 and 91,000. In downtown the values are higher, ranging from Can$127,000 to 202,000. Once again, Cap-Blanc shows the lowest housing values averaging between Can$ 44,000 and 65,000.

The following socio-economic profile shows that middle class families are owners of their residences in Cap-blanc and that wealthier and older residents own their homes within the low city, around Place Royale and in the Vieux-Québec Port. These older residents are unmarried without children.

Poorest families rent housing in the uptown sector.

A majority of male bachelors occupy a mixed zone of rental or private residences, in the section delimited by Hotel-Dieu Hospital, Saint-Jean Street, the

Coast of the Factory, a portion of Saint-Louis street and the circumference of the Frontenac Castle. A last more heterogeneous zone congregates tenants, of male or female bachelors, from the Citadel to Auteuil Street.

The distribution of household incomes confirms the dichotomy between downtown and uptown. Estimated average income in uptown is Can$18,050, whereas that of downtown is Can$41,843. A strong proportion of students explains these low incomes. The unemployment rate is more significant in the uptown sector than in downtown. We observe a constant rate of 14.6% in uptown, in 1996 as in 1990 and of 4% in downtown, especially in the Place Royale sector. The average for Quebec City was 10.4% in 1996. Therefore, we can read the uptown low incomes as the result of chronic unemployment. Moreover, tenants occupy 83.9% of the private residences, whereas downtown it is estimated that 49% are owner-occupied.

Michel Simard noted Vieux-Québec population's uniqueness in these terms:

> The Vieux-Québec seems a male majority district, inhabited by adults from 20 to 39 years, more English-speaking, more mobile, more educated , working in public administration and professional services, counting more self-employed workers and tenants. The Vieux-Québec Population is thus definitely different from the general profile in the Quebec area (2002, p. 4)

This distinctive society described by Simard calls for comments. First, the strong Anglophone minority must be seriously taken into account. Indeed, the English-speaking population accounts for no more than 0.3% of the population of the large town of Quebec where 671,889 people lived in 1996. However in the Vieux-Québec they represent 6.2% of the residents. We have to keep in mind that in the Province of Quebec 9% of the population are Anglophone and 8% Allophone living mainly in greater Montreal.

We know that before the *Revolution Tranquille*, in 1956, the Quebec Anglophones represented 17% of the total population of the Province or 691, 438 people, (which is less than the population of Quebec City in 1996). However, it seems difficult to find accurate statistics of the Anglophone presence in the Vieux-Québec for 1956. We simply suppose that, in spite of the departure of significant numbers of English-speakers from the Province of Quebec, a strong feeling of veneration still connects the Vieux-Québec to the Anglophone population. This feeling of Anglophone cultural identity and sense of place was not taken into account in these two surveys, which statistically undermine the Anglophone presence and thus their attachment for the historic district[5].

Second, the prevalence of males leads us to believe that the Vieux-Québec resembles more and more the Vieux-Carré in New Orleans where the Gay population tends to become the majority. It would be interesting to prove whether there is such a trend in Vieux Québec. This educated population belongs to a community that prizes heritage places and invests in gentrification and renovation. They have fought for the recognition of urban historic places in North America in order to save them from destruction.

There is great difficulty in attracting and maintaining residents in the old center of Quebec: In Vieux-Québec very few children are present. 24.5% of the population in Quebec city are young people from 0 to 19 years old but only 8% in Vieux-Québec and only 4.6% in the downtown part. There too, old centers appear as dead places that lead to the fossilization of the district even to a ghettoïsation of these spaces. They become places for adults, more generally bachelors, who are subjected to temporary flows of tourists whose numbers have increased since the birth of mass tourism.

Lussault points out that an image, strictly speaking, does not exist. We are facing a complex textual or iconic representations structure of the heritage city (1993, p. 351). Vieux-Québec images depend on the eyes of the beholders. For the Vieux-Québec Tourism Office and the Quebec Urban Community Congress tourist bureau, the Heritage center is a simple "marketing product" (OTCCUQ, 1999: 15). This appreciation resembles that underlined by the president of the Traders Union, namely that of a vast open-air shopping centre. However Vieux-Québec residents feel proud of their heritage environment and they choose to live and stay in Vieux-Québec. Louis Germain was the official journalist of the Vieux-Québec Citizens gazette in 1991

> " We are the living of Vieux-Québec and we consciously, freely, and voluntarily chose to
> reside in this place. Why? Because we do love this historic place. In addition to this
> love, we all want to appropriate it, defend it, organize it, promote it, embellish it, and
> show it ". (Author's translation). (Germain, 1991, p. 6).

This act of faith reveals to us the strong feelings invested in this area and its venerated image. Maranda (2001) underlined two major Vieux-Québec life attractions, which are the historical character and the beauty of the site. The official advertising slogans praise the city using these same two characteristics mentioned above.

If in 1990 tourists were appreciated in Vieux-Québec, (only 7% of the answers show tourists as a disadvantage), ten years later they represent a nuisance to their way of life. Thus 43% of the respondents complain about the noise and 36% identify the tourists as a source of embarrassment. For 46.3% of the respondents, these tourists strongly hinder or hinder just enough the district life. They identify the cars' and buses' harmful effects as a major disadvantage to their city. Parking has also become more and more difficult. Answers also reveal that tourists' overconsumption is harmful to their tranquil environment. However Vieux-Québec residents retain a very strong attachment to their urban setting. This strong feeling is based on a general appreciation of the architecture of the built environment, housing and public places and of the maintenance of the district by the city. The architectural heritage accounts for 94.4% in the public's general appreciation.

The second pole of the public's general satisfaction is determined by the high quality of public services related to the cultural activities (security, health and educational services). The answers also reflect the pride of living in an internationally famous historical place. Respondents' pride encourages them to assume more responsibility with regard to the protection of the neighborhood. The beauty and the historic character of the site re-emerge as major attractions for staying in Vieux-Québec in 1990 as in 2000. The confirmation of the positive

Figure 5. Caleches and Tourists in Upper town, Michel Delisle 2003

attractivity of the built environment and district life is more evident that decade between 1990 and 2000.

It seems clear that the Vieux-Québec residents kept on living within the limits of the district ten years later. In more than 70 % of the cases, residents frequently walk in the Vieux-Québec streets (88.7% of respondents in 2000 against 77.3% in 1990). Many restaurants of the old city do not serve just the tourists but also the residents of the neighborhood who regularly use them (89.2%). We can explain the exceptional frequentation of these restaurants by the high percentage of the unmarried without children, who reside in the old district.

We remark a great change among the residents of Vieux-Québec due to the influx of visitors. In 1990, 44.6% of the residents accommodated outside guests compared to the high rate of 69.9% in 2000. These percentages indicate that the residents are proud to receive visitors who find the area attractive. We must take into account the length of the stay of the respondents. In 1990, the average residence period was 8 years; in the following ten years the stay is shortened to 6.2 years. Nevertheless, 25.5 % of the respondents had lived in the historic center for at least ten years and 5.6% had occupied these places since 1980. A significant proportion of 25.7 % of respondents returned to reside in the Vieux-Québec after already having left once. Thus the surveys show a true attachment of some of the residents to Vieux-Québec as their place in life.

THE CITY'S MARKETING DISCOURSES ON THE WEB.

Advertising generates an imaginary landscape (Cowen, 2002; Gregory, 1994). Advertisements, in our era of globalization, produce images for marketed cities based on an interpreted culture. In a broad sense, this includes political elitist culture and sub-cultures. On the Internet advertising for the virtual Vieux-Québec reinforces common ideas supplied by the official rhetoric, which are the French character of the place, the authentic World Heritage, and the participation in an adventurous experience through History.

Museums are the first to form part in this discourse. For instance the Séminaire de Québec (Old Seminary Museum) displays this advertisement on its web site in English:

> Discover this jewel of our heritage. Travel back in time (...) Enter the history of this great human adventure! (...) The adventures of French speakers in America. (...) Discover the history of French America[6].

Musée de la civilisation (the Civilization Museum) and Musée de l'Amérique Française (French America Museum) both celebrate this uniqueness in the heart of Vieux-Québec. We can read on French America website:

> Its programming is an invitation to relive the great adventure of the French people in America, to look into their every move, to do well on their way of life, to understand the spirit that was their driving force. A full grasp of this heritage of the modern world, heritage that bears witness to the determination and dynamism of millions of peolpe, that sheds light on the present and the future of every Quebecer, both individually and collectively[7].

> Musée de la Civilisation, a theme Museum, coordinates various centres such as Urban Life Interpretation Centre and Place Royale Interpretation Centre. We can read in English:

> The Musée de la civilisation links the past, the present and the future. While remaining strongly rooted in the Québec reality, it projects a new, attentive and dynamic outlook on the human experience in its whole, and on civilizations from the world over[8].

The Place Royale Interpretation Centre «presents the history of this site, considered the cradle of French civilization in the Americas. (...) From its origin as a fur trading post to its present status as Québec's provincial capital, Québec City has restored and celebrates elements from its past »[9].

On the same web site we can read:

> Québec City has embraced its history and keeps it alive.

> Here, you will find no less than 32 museums, exhibition halls and interpretation centers that highlight the heritage of the past and the energy of contemporary artists and craftspeople. A visit to one of these museums is like walking down a path leading to the discovery of history, art, heritage, and culture.

On the Quebec website we can find the City Hall narrative, which emphasizes in its presentation the physical landscape dichotomy and a glorified "postcard" image of the capital of Quebec.

Figure 6. Place Royal Interpretation center, Martine Geronimi 2003

Quebec City: An Historic Treasure

The only fortified city in North America, and from the top of it you can admire the maze of narrow, winding streets and sloping roofs. Perched atop Cap Diamant, surveying the St. Lawrence River, Québec City is one of the landmarks of North American history. Included on UNESCO's prestigious World Heritage List, Québec City has retained its European atmosphere completely. (…) Québec City, the cradle of French civilization in North America, is today a busy seaport, an important centre of services and research, a cultural hot spot and, of course, the provincial capital. Québec City is located 250 km east of Montréal and about 850 km north of New York City[10]. (Not translated)

This site, especially created for American tourists, shows in its web page a reassuring image as to the usage of the two languages

Québec City is almost entirely French in feeling, in sprit, and in language; 95% of the population is Francophone. There is an existing tension and an ongoing battle in the province of Quebec between the Francophone and Anglophone communities. Even though the English language is used everywhere as the main means of expression for the tourism industry around the world, in Quebec the tension is amplified by the constant use of the English language by the tourism industry personnel of the city. However in 2000, before the city mergers, about 5% of the 648,000 citizens of Quebec City had English as their mother tongue, this is excluding Francophones who work in hotels, restaurants, and shops where they deal with Anglophone tourists every day.

Figure 7. Porte Saint-Louis, Upper Town, M. Delisle 2003

This historic treasure presents an Upper Town and a Lower Town. The Website insists on the old strategically military vocation of the Upper town and the maritime and commercial elements of the Lower Town.

> It is at the top of this natural citadel that Samuel de Champlain established in 1608 a fur trading post. The district had from the very start of the colony a military and administrative character, the strategic heights of the Cap Diamant determined its vocation. The Upper Town is inhabited by the soldiers, the civil servants, and the members of the clergy[11].

For the Lower Town we can learn:

> The Vieux-Port de Québec (old port), at the confluence of Saint-Charles River and the St.Lawrence River, is literally steeped in history and provides a window on the maritime activity of yesteryear. The lock maintains constant water level within basin Louise that gives access to a marina that can accommodate several hundred boats.

> The Marché du Vieux-Port (market), where farmers from the region sell their produce. (sic)

> The rue Saint-Paul, where well-known antique shops have been in business for several years. You will find various boutiques, art galleries, restaurants and cafés where time seems to have stopped.

The description for American tourists of the Lower town emphasizes the unusual characteristics of the harbor. It doubles the Lower Town interest for the American or Anglophone tourists as it insists on Past and Modernity, as done for the Port and for Place Royale. The restaurants, coffee shops, and art galleries bring modernity and a form of familiarity to the international tourist in a Francophone heritage place. This web site demonstrates that Quebec and French Canadians are not just relics of the past. They are part of Modern American Life.

> This is one of the oldest districts in North America. Visitors can admire three centuries of history as they stroll down its narrow, colorful streets. Nestled between cap Diamant and the St. Lawrence River, Place Royale has now been almost completely restored and is reserved for pedestrian traffic[12].

Specifically Place Royale appears to be a common place and a sort of reserve for tourists. This superficial description of the restored district minimizes the importance that Quebeckers give the place as the Cradle of the Francophones in America. Place Royale is a focus point for the Francophones and requires our attention and explanation.

> Place Royale
>
> Birthplace of French America, the historical site of Place-Royale welcomes thousands of visitors each year. Its narrow and picturesque streets have witnessed three centuries of history. Some of its buildings are a component of the cultural complex managed by the Musée de la Civilisation, responsible for the site's animation and interpretation [13]

The official City Hall Website adds in French:

> Around Place Royale, the houses are built of heavy stones and are protectively enclosed behind the walls. High chimneys adorn the roofs and the windows contain small rectangular window-panes. These are just as essential as the historical testimonies of the small colony, which inhabited it. This represents a certain manner of building and of living[4]. (Author's translation)

If we compare this description with the portrait done by Lonely Planet, you can note a major difference:

> The Lower Town's main meeting point is the 400-year-old Place Royale, a human vortex surrounded by nicely aged buildings and overpriced tourist emporiums.
>
> On the western side of Place Royale is the Église Notre Dame des Victoires, a modest edifice built in 1688 and ranked the oldest stone church on the continent.(..[15])

This quote begs for a comment since the Lonely Planet addresses mostly backpackers, people who believe they visit more closely sites with a difference than the majority of tourists who are superficial visitors (Edensor, 1998) providing yet another example of the tension previously mentioned.

UNESCO's World Heritage List endows this historical site with legitimacy; it provides a chance to create attention, and thus the label becomes a brand name. This award becomes for tourists a guarantee of high quality and of authenticity. In 1985 the justification for the inscription to the world heritage list was:

> Québec was founded by the French explorer Champlain in the early 17th century. It is the only North American city to have preserved its ramparts, together with the numerous bastions, gates, and defensive works, which still surround Old Québec. The

Upper Town, built on the cliff, has remained the religious and administrative centre, with its churches, convents and other monuments like the Dauphine Redoubt, the Citadel and Château Frontenac. Together with the Lower Town and its ancient districts, it forms an urban ensemble, which is one of the best examples of a fortified colonial city[16].

Figure 8. Lower town, Petit Champlain Street. Martine Geronimi 2003

This Unesco listed urban ensemble is an undisputed example representative of a globalised world (Meethan, 2001; Logan, 2002). It seems to conduct Vieux-Québec to the commodification of its virtuous French Past identity. That means that the inner core of the city has become a cultural heritage theme park featuring French Folklore. The marketing of Québec City has showcased the French Town and thereby also pinpointed the first colonial period. This major tourist attraction is embedded in a double bind discourse in which the collective identity is explained to the provincial visitors as a narrative that produces a strong feeling of uniqueness compared to the Anglophones' discourses for the rest of Canada. Place Royale is the key symbol of the ideology for an independent Quebec state.

Place-royale is viewed as the cradle of French-speaking civilization in North America, and not as the cradle of the two founding societies of Canada. Place Royale is a place of commemoration of the French-speaking origins in America and a monument devoted to the quest for Quebeckers' independent identity (Noppen and Morisset, 1998). The semantical slip is interesting: this cradle bears in its center the mythical speeches of the origins. The receptacle contains two complementary discursive concepts that are associated with the memory ritual - the duty of memory-and the dynamics of the testimonial identity crisis of Quebeckers in the recent past. The ideological jump also shows in geographic dynamics: from a speech of federal-provincial consensus, we jump to a purely Quebecker reappropriation of the endorsements. By the same token, Nationalists set up in this place a symbol of the fight for Quebecker identity.

The forms induced by the narrative show that, at the beginning of the project, Place Royale was an empty vessel whose architecture was incompletely reconstructed in the French style. Nowadays, the complete Quebecker territorial model would be anchored in this vessel, which, in fact, would also be filled with political meanings. These abstract values, associated with the Quebecker identity, are presented to international tourists, and especially to the provincial tourists through the Center of Interpretation of the Place Royale. Through this didactic position with a post-modern taste, Place Royale becomes a symbol of real legitimacy: the concrete and virtuous demonstration of a collective destiny. This discourse in images, those of an architectural landscape completed by an up-to-date website, is allied with the essential remarks of the official rhetoric. It shows to the tourists, who do not have an extensive knowledge of Quebec's political strategies anyways, the face that Quebeckers would like to remember with pride, admiration or pleasure.

However, a new incontrollable dynamic is applied to the visited places: the glance of the Other. This takeover of the proposed territorial ideological model by the tourists' gaze creates a new perspective on the place that is not easy to decrypt. Moreover, we have to take into account this double dynamic before speaking of Disneyfication or theming of the landscape. From now on, the Place Royale's touristic success depends on this new perception. The imaginary of the post-modern travelers transforms vernacular landscapes and thus recomposes and internalizes them according to "softer" codes composed with simplistic international values but also with a rich individual capacity for sublimation and for enchantment.

Moreover, Quebec's symbolic economy is also present on the Internet through the preparation of its 400th anniversary, in July 2008. This major event crystallizes the French heritage discourse and *independentist* narrative. Quebec City is presented as the National Capital. Quebec City is celebrated as the Cradle of the French civilization in North America and the Founding of French North America, a French empire called Nouvelle France[17].

This celebration seems to be not only a moment of remembrance for the Event's organizers but also:

> «an opportunity to involve the public over the next few years in an exciting and stimulating undertaking. It is also a unique and powerful way to draw the community [the Quebeckers] together around a single project and spur development throughout the region»[18].

The "Quebec400" official web site validates the idea that Quebec's symbolic economy is a stimulator for any undertaking, especially economic undertakings that come in the form of commoditization of cultural artefacts to attract international tourists. This shows that Quebec is a modern nation within North America while its national capital is an authentic French heritage place in a non-existing independent country. Yet, obvious oppositions continue to exist between the traditional French past and the modern American way of life for contemporary Quebeckers:

> Any number of themes could be considered, such as economic development, a celebration of history, fields of excellence, culture and heritage, high tech, or a public festival. Although the historical aspect of the event will remain important, Quebec and its national capital must be presented as something more than a living museum of history; they must be seen as a vibrant community celebrating an entire past, present, future continuum to the tourists and prospective investors.

Organizers insist on the distinctive and attractive qualities of Quebec City and conclude:

> Québec City is not a museum frozen in time. On the contrary-it's a modern, people-friendly city with its sights resolutely set on the future.

Investing in Vieux Québec' image

Besides the official virtuous narrative and beyond the touristic discourse, Vieux-Québec exists as an aesthetic marketing brand. Anglophones in Canada for instance, also appreciate Vieux-Québec. The Anglophone Business sector does re-appropriate the imaginary of the city in a commercial way. This is different from the direct link between heritage and tourism. This form of symbolic economy is obvious on the Internet through the example of Old *Quebec Street Shoppes and Office Suites*[19].

> What is unique about Old Quebec Street is the vision. The mandate of the developer, Barrel Works Guelph Ltd. - a joint venture between Kiwi Newton Construction Ltd., J. Lammer Developments Ltd., and Terra View Homes - was to create a European Streetscape that appeared to have been built years ago. To do this, major fundamental changes had to be designed into Old Quebec Street.

The developer's advertisement plays on the uniqueness of the place, its European flavor and the (re)creation of a fake landscape which appears to be very old. The promoter is reappropriating the already authentically sanctioned Quebec past in a luxurious and aesthetic old Downtown in Ontario. These business people are selling the Quebec name in another historical downtown, that of Guelph. This fake downtown is also destined to become a large commercial centre with more than 500 retail shops.

The developers state that:

> Exceptional architecture honouring Guelph's historic past is the focus of the team responsible for creating this world class redevelopment. Once complete, Guelph will have a jewel in its crown like no other city. Everybody, residents and visitors alike, will share new fond memories as they work, shop and play in Guelph's downtown.

This is the key symbolic phrase: Placing the Jewel Back into the Crown. This commercial reappropriation is also a political one. Old Quebec was in the 19th century the old capital of Canada. Promoters are playing on the tempting and unrealistic idea of reappropriation of the political past, a nod in direction of the Quebecker separatist movement. As Lears assumes, «advertisements urge people to buy goods, they also signify a certain vision of the good life; they validate a way of being in the world. They focus private fantasy; they sanction or subvert existing structures of economic and political power. Their significance depends on their cultural setting (1994, p. 1). Advertisements from the Guelph developers also put forward an image of abundance and possible entertainment, while they maintain that modernity plays with the past.

> Located off St. George's Square, the recreation of a covered Old Quebec Street will take full advantage of shoppers seeking a destination where they can find entertainment and a unique blend of specialty stores.

> Old Quebec Street in the heart of downtown Guelph offers quality stores and office suites in a unique European style streetscape where natural sunlight cascades down from a retractable skylighting towering overhead.

CONCLUSION

The Vieux-Québec, in Quebec City, has become to international tourists a high-end commodity since its attribution of the Unesco World Heritage label in 1985. This socially constructed area carries French symbolic values, values that are confirmed by its language, culture, and history. World heritage landscapes, as those of Vieux-Québec, are cultural landscapes in which these very same symbols and images of the past are recurrent. In our era of globalization, commodification of Old-Quebec's culture is embedded in the symbolic landscape which is presented to the mass of tourists that visit the city each year. Quebec City has also found a new way to strengthen its reputation as a successful place for tourists through the Internet, presenting its best asset to the world: a historical yet modern city of North America.

In this chapter we have tentatively demonstrated that Old-Quebec is a perfect example of the collapse of barriers in space and time and of a progressive loss of exclusivity. The official proclamation of the French political identity of the old district is reappropriated by tourism, its residents and various business interests.

Since 1985, when the world Heritage city label conferred additional prestige on Vieux-Québec, the city has seen an increase in the number of its international customers. Specifically Place Royale appears to be a common place while, at the same time, a sort of reserve for tourists. However surprising it may seem, we must admit that Vieux-Québec residents retain a very strong attachment to their urban setting; they feel proud of their heritage environment and they choose to live and stay in Vieux-Québec. Even though the Vieux-Québec is a 'no families' area, it is now repopulating thanks to the efforts of the city's municipal government.

Nevertheless, the Vieux-Québec images depend on the eyes of the beholders. For the Vieux-Québec Tourism Office and the Quebec Urban Community Congress tourist bureau, the Heritage center is a simple "marketing product". It does lead Vieux-Québec to the commodification of its virtuous French Past identity.

The marketing of Québec City has provided a showcase for the French Town and its first colonial period. Place Royale is the key symbol of the ideology for an independent Quebec state. Quebec City presents itself as the National Capital on its Internet website. Quebec City is renowned as the Cradle of the French civilization in North America and the Founding of French North America, a French empire called Nouvelle France. It is fascinating to note that even though Québec City is celebrated for its historical values, against all potential detractors, it is presented as not being a sort of museum frozen in time.

Its symbolic economy is not only discernible on the Quebec web site but is also virtually present on the World Wide Web. A significant example of this is Guelph's Old Quebec Street Shoppes and Office Suites. In this other purely Anglophone Canadian city, the re-appropriation of Old Quebec does not refer to the French past at all. It seems to reveal our postmodern world with a twist of irony and refer to an aesthetic marketing brand. It is amazing, from the French point of view, to see that Anglophone promoters play on the tempting and probably unrealistic idea of reappropriating the political Canadian past. Old Quebec is also a strong reference point for Anglophones in Canada, making it a tempting money-making plans by creating a play on its political past and present.

We may conclude that the promotion of Old Quebec, as of every heritage city all around the world, is subordinated to the growing use of the web. As Turtinen states:

> "This is why the use of media is crucial as means of representing World Heritage. It is through use of media, through technologies of visualisation and textualisation that World Heritage is brought to life. The products are media for making World Heritage visible and intelligible. Furthermore, media makes the necessary spatiotemporal mergers possible. Media render, with all the immediacy of visual images and informative texts, the peculiar co-ordination of space, time, and people (territory, history, and society) that makes humankind an identifiable kind of imagined and moral community." (2000, p. 20)

Finally, Old Quebec has become a local and regional hot spot for publicity, a taste of French cultural identity and of course, tourism. The example of Old Quebec demonstrates the valorisation of heritage as cultural capital and furthermore as a symbolic good in our Society of Spectacle.

Year	Quebec City	Old Quebec	percentage	Old Quebec Cap blanc	percentage
1941	150 757	10 996	7,3%		
1961	171 979	10 252	6,0%		
1981	165 968	4 083	2,5%		
1986	165 580	4 383	2,6%	5 179	3,1%
1991	167 972	4 450	2,6%	5 180	3,1%
1996	167 634	4 607	2,7%	5 297	3,2%

Table 1. Residents of Old Quebec
Population of census Quebec metropolitan in 200, p. 696.400

Table 2 Growth Old Quebec Population 1981-1996

Year	Upper Town	Lower Town	Cap Blanc	Total
1981	3 453	630	843	4 926
1986	3 525	858	796	5 179
1991	3 215	1 235	730	5 180
1996	3 185	1 422	690	5 297

Source: Statistics Canada, Census of Canada, 1971-1996

Department of Geography
University of Quebec in Montreal
Special thanks to Marie-Évangeline Pouyer for her English revision

NOTES

[1]http://www.copa.qc.ca/forces/anglais/article3.html

[2]The Quiet Revolution is the name given to a period of Quebec history extending from 1960 to 1966 and corresponding to the tenure of office of the Liberal Party of Jean Lesage. The term appears to have been coined by a Toronto journalist who, upon witnessing the many and far reaching changes taking place in Quebec, declared that what was happening was nothing short of a revolution, albeit a quiet one. Bélanger Claude on : http://www2.marianopolis.edu/quebechistory/events/quiet.htm

[3]Already in 1981, a study was published on the topic, entitled Quebec, une ville à vendre (Quebec a city for sale)

[4]Tables at the end of the chapter,

[5]http://www.ocol-clo.gc.ca/archives/sst_es/2004/jedwab/jedwab_2004_e.htm#I_c

[6]http://www.mcq.org/maf/aaindex.html

[7]http://www.mcq.org/maf/aamaf.html

[8]http://www.mcq.org/mcq/aamcq.html

[9]idem

[10]http://www.telegraphe.com/quebec/QuebecCity.html

[11]http://www.telegraphe.com/quebec/vieuxquebecin/indexen.html

[12]http://www.telegraphe.com/quebec/vieuxquebecex/indexen.html

[13]http://www.telegraphe.com/sites/indexen.html

[14]http://www.ville.quebec.qc.ca/fr/ma_ville/quartiers/portrait/portrait_vieux-quebecbv.shtml

[15]http://www.lonelyplanet.com/destinations/north_america/quebec_city/attractions.htm

[16]http://whc.unesco.org/sites/300.htm

[17]www.Quebec400.qc.ca

[18]http://www.quebec400.qc.ca/ang/actualites/index.asp

[19]http://www.guelphdowntown.com/oldquebecstreet/

REFERENCES

Auge, M. (1995) Non-Places: Introduction to an Anthropology of Supermodernity. London: Verso.

Baudrillard (1981) For a critique of the political economy of the sign. St. Louis, MO: Telos Press

Benjamin W. (1982, 1928) "The Work of Art in the Age of Mechanical Reproduction," in Illuminations, Engl. trans., London, Fontana.

Bouchard (1996) Entre l'Ancien et le Nouveau Monde: Le Québec comme population neuve et culture fondatrice, Ottawa, Les Presses de l'Université d'Ottawa.

Bourdieu P. and Coleman J.S. eds (1991) Social Theory for a Changing Society, Coleman; Russell Sage Foundation.

Castells, M. (1996). The Information Age. Economy, Society and Culture. Volume I: The Rise of the Network Society. Oxford: Blackwell

Cox, K (1997) "Globalization and geographies of workers' struggle in the late twentieth century.", The Geographies of Economies, Ed. Lee, R and Wills, J, London: Arnold, pp 177-185.

Courville, S. (2000) Le Québec, genèses et mutations du territoire. Synthèse de géographie historique, Sainte-Foy, Les Presses de l'Université Laval

Cowen, T (2002) Creative Destruction: How Globalization is Changing the World's Cultures. Princeton University Press, Princeton NJ

Dann,G. M.S. ed (2002) The tourist as a metaphor of the Social World, Wallingford CABI Publishing

Debord, G. (1967) La société du spectacle . Third edition, 1971. Paris, Éditions Champs libres.

Debray, R. (1992) Vie et mort de l'image : une histoire du regard en Occident. Paris, Gallimard

Debray, R. (1998) Trace, forme ou message ? Cahiers de Médiologie 7, n° spécial "La confusion des monuments".

Gregory, D. (1994) Geographical Imaginations. Oxford: Blackwell.

Dicken, P, Peck, J and Tickell, A (1997) "Unpacking the global." Geographies of Economies, Ed. Lee, R and Wills, J, London: Arnold, pp 158-166.

Edensor, T. (1998) Tourists at the Taj: Performance and Meaning at a Symbolic Site. London: Routledge.

Ezor Q. (1981), Une ville à vendre Montréal, Éditions coopératives Albert Saint-Martin

Florida, R. (2002): Bohemia and economic geography. In: Journal of Economic Geography 2 (2002) pp. 55-71

Geronimi, M. (1996) Le Vieux-Québec au passé indéfini. Entre Patrimoine et Tourisme, Sainte-Foy, Université Laval, département de géographie, mémoire de maîtrise.

Geronimi, M. (1999) Permanence paysagère et consommation touristique : Le cas du Vieux-Québec In Beaudet G., Cazelais N. et Nadeau R (dir) (1999) L'espace touristique. Québec, Presses Universitaires de Montréal, pp.?

Geronimi, M (2001) Imaginaires français en Amérique du Nord. Géographie comparative des paysages patrimoniaux et touristiques du Vieux-Québec et du Vieux Carré à la Nouvelle-Orléans, Québec, Département de Géographie, Thèse de doctorat.

Geronimi, M (2003) Québec et La Nouvelle-Orléans, paysages imaginaires français en Amérique du Nord, Paris, Belin.

Lamarche, J. (1998) Comment faire une ville ? Murs Murs, 9 (1), pp. 3-5

Lamarche J. and Simard M. (2002) Vivre dans le Vieux-Québec, rapport de recherche, CCVQ

Lears J. (1994) Fables of Abundance: a Cultural History of Advertising in America, New York: Basic Books

Lee, R. and Willis J. (1997) Introduction? In Geographies of Economies. London, Arnold, pp.?

Leyshon, A (1997) "True stories? Global dreams, global nightmares, and writing globalisation.", Geographies of Economies, Ed. Lee, R and Wills, J, London: Arnold, pp. 133-146.

Logan, W (2002) The Disappearing 'Asian' City Protecting Asia's Urban Heritage in a Globalizing World, Oxford University Press (China), Hong Kong

Lury, C. (1996) Consumer Culture. New Brunswick, New Jersey: Rutgers University Press

Lussault, M. (1997) Des récits et des lieux : le registre identitaire dans l'action urbaine. Annales de géographie, 597, pp. 522-530.

Lussault, M. (1993) Tours : images de la ville et politique urbaine. Tours : Maison des sciences de la ville, Université François Rabelais.

Maranda, P. (1991) Qui sont les citoyens du Vieux-Québec? Comment ressent-on la vie dans une ville du patrimoine mondial? Rapport de Recherche, Département d'Anthropologie, Université Laval.

Meethan, K. (2001) Tourism in global society: place, culture, consumption. Basingstoke, Palgrave.

Morgan N. & Prichard A. (1998) Tourism Promotion and Power Creating Images, Creating Identities, John Wiley and Sons Ltd

Morgan N. & Prichard A. (2001) Advertising In Tourism And Leisure. Oxford ; Boston : Butterworth-Heinemann

Noppen L. & Morisset L. (1998) Québec de roc et de pierres. La capitale en architecture. Québec, éditions MultiMondes

Otccuq, Plan de développement-marketing de la région touristique de Québec 1998-2002- Mise à jour 1999. Québec

Ppideaux, B. (2002) "The Cybertourist", in The tourist as a Metaphor of the Social World, Ed Graham M.S. Dann, publisher, pp.?

Rocher, G. (1973) Le Québec en mutation. Montréal: Les Éditions Hurtubise HMH ltée.

Sack, R - D. (1997). Homo Geographicus: A Framework for Action, Awareness, and Moral Concern. Baltimore: Johns Hopkins.

Scott, A.-J. (1997) The cultural economy of cities, International Journal of Urban and Regional Research, p. 21, p. 2, pp. 323-39.

Scott, A.- J. (2000) The Cultural Economy of Cities. Sage, London

Terkenli, T. S. (1992) "Landscapes of tourism: towards a global cultural economy of space?", Tourism Geographies , Volume 4, Number 3, pp. 227 - 254

Turtinen J. (2000) Globalising Heritage – On UNESCO and the Transnational Construction of a World Heritage, SCORE Rapport serie 12

Zukin Sh. (1995) The Cultures of cities, New York, Blackwell Publishers

Official discourses on the web

Forces Journal paper

http://www.copa.qc.ca/forces/anglais/article3.html

Bonjour Québec for tourists

http://www.bonjourquebec.com/anglais/idees_vac/villes/quebec.html

Museum for tourists and curious

http://www.mcq.org/maf/aaindex.html

http://www.ovpm.org/ville.asp?v=185 UNESCO

Commission de la Capitale nationale

http://www.capitale.gouv.qc.ca/souvenir/monuments/default.html

Other discourses on the web

http://www.canada.co.nz/media/story_35.htm

Quebec Immigration discourse

http://www.immigration-quebec.gouv.qc.ca/anglais/avantages quebec/choose_quebec_intro.html

City Hall narratives

http://www.ville.quebec.qc.ca/fr/ma_ville/quartiers/portrait/portrait_vieux-quebecbv.shtml

http://www.ville.quebec.qc.ca/fr/ma_ville/quartiers/portrait/portrait_vieux-quebechv.shtml

Old Quebec citizen group

http://membres.lycos.fr/citoyenvieuxqueb/Memoire/m_culture_02.htm

A-M. D'HAUTESERRE & T. TERKENLI

TOWARDS REWORLDMENT: CONCLUSIONS

Paris Disneyland: a dazzling but incongruous landscape of entertainment, source A.-M. d'Hauteserre

Landscapes are the settings of interaction, the context for economic activities. Today's landscapes translate the more recent ways capitalism, as a social relation, through new strategies of desire and at different scales, seeks to turn a profit. The purpose of this volume was to identify trends and structures articulating the new cultural economy, as well as to understand how these new dynamics of production, circulation and consumption differentiate spaces and places. In its more metaphorical sense, the term 'new cultural economy of space' employed here was intended to embrace and make an attempt towards interpretation of spatial change through a cultural lense, broadly defined. The outcome is a variety of conceptual and analytical approaches to the subject and a highly diverse mix of empirical data sources, illustrative of the tools and techniques necessary to document and understand processes of the 'new cultural economy of space'. In this way, this

T.S. Terkenli & A-M. d'Hauteserre (eds.), Landscapes of a New Cultural Economy of Space, 239-245
© *2006 Springer. Printed in the Netherlands.*

volume illustrates different ways of interpreting spatial change, but all authors seem to think of them as more or less contingent on a cultural negotiation of space and landscape.

Landscapes exist not just in space but through time and cannot be detached from their historical context. All landscape analyses in this volume, as well as the following discussion, rest on this premise. For instance, authors in this volume have had to come to grips with ephemeral and disposable or staged landscapes resulting from processes of unworldment, most common examples of which exist in tourist destinations where they erase the local historico/cultural layerings. Such juxtapositions are justified as aids to 'economic growth', generally applicable to the growth of the economic assets of investors (creators of fictitious commercial landscapes) rather than to any local indigenous 'progress' (d'Hauteserre, Geronimi). Movement and interconnectivity have broken down not just spatial but also temporal barriers so that landscapes no longer identify particular socio-economic systems. Bataille had already shown in the 1930s how enchantment and obsession propelled accumulation. In order to resist more effectively the consequent dislocation and non-engagement of artificial 'non-places', the trajectory of 'commodification' of these artificial worlds needs to be deconstructed as a historical case to understand how we can all be seduced by the construction of homogenised tastes.

Rather than effecting a cultural analysis of market practices or demonstrating the consequences of the appropriation of matters cultural for economic advancement, authors endeavoured to identify how contemporary landscapes express various hybrid and temporary coalitions or entanglements in everyday life, recognising that the local is never purely local. Their approach emphasises the variety of actors contributing to this 'new cultural economy of space' at all different geographic scales from the individual (body) to large investors, but also in relation with other kinds of actors (humans as well as natural elements and material devices), as identified by Latour (1991) and Callon (1991). Moreover, the presence now of various 'others' in the midst of previously exclusive enclaves of the centre (Massey, 1994) creates contexts of extraordinary ethnic, religious and linguistic diversity which further stimulate processes of the new cultural economy of space.

Theoretical accounts in geography and the other social sciences have generally undervalued the 'everyday' and the intimate, considered until now as 'just incidental'. Research has to matter to lay people, but those who use the results have felt that culture has been presented, at the theoretical level, as a series of vacuous abstract concepts (like Bourdieu's *habitus*) in the attempt to define this quotidian. Its embodied practice at the level of individuals, on the other hand, reveals complexities, contradictions, divergent tendencies and the tensions that result from individual performance. The body is, after all, the scale at which one makes it from one day (hour?) to the next. The individual scale includes the required bodily exertion, the wear and tear of individual workers. The study of embodied performance overcomes the analytic erasure of daily life that occurs through abstraction from smaller to larger scales.

Authors also describe the intersections and encounters of practices at and between different scales, though it must be noted that individuals, not abstract entities, are always involved in these mutually constitutive interactions, whatever the scale or the size of the organisation they belong to. Crouch, for example, underlines how the notion of (individual) performance needs to be considered in relation to socio-cultural formations of accumulation. Crang (1997) confirms that "networks of *interpersonal* relations multiply scale embeddings of economic practice and organisation" (our emphasis). We thus need to turn our attention to individual embodied struggles in today's cultural economy, because daily life and the inscription of work in bodies are often obfuscated when attempting to develop a 'super organic' understanding through generalisations at larger scales.

Study of the individual scale facilitates the integration of encounters across scales, including global intrusions on local scenes: "…[the individual] embodies various forms of capital… possesses the ability to mediate and interpret between the areas of culture and of service provision … closes gaps in the urban with new social, entrepreneurial spatial practice [and] … is a communicative provider of transfer services between sub-systems" (Lange, p. 40). The outcome of encounters along links or within networks can increase or decrease an individual's ability to perform. Crouch (p. 20) confirms that "…acts of making of space *through everyday actions* are mutually engaged in multiple human flows" (our emphasis). Reference to the scale of everyday life is a constant theme of this volume.

Cultural transactions are but one of the organisational arrangements of this new cultural economy which relies on the circulation of information and knowledge, on social regulations and infrastructural investment, 'as the cultural and the economic move in lockstep' (Sheller & Urry, 2000). The consequent aestheticisation of the social has led to an emphasis on appearance, display and the management of impressions, so that performance in certain economic activities has become an occupational resource (Lange). As 'impressions have now often taken over the technical aspects of labour', performativity of economies is now mutually interdependent with performance of identity (Adkins & Lury, 1999, p. 599).

Authors in this volume emphasize how no production process can be addressed through the study of just one kind of material and/or mean of production. They understand that economies are not simply bound by the rules of profit and loss. They have avoided, though, the danger of uncritically celebrating the cultural embeddedness of the economy, which often displaces social relations or 'cultivated dispositions to act', such as today's growth of a more individualistic 'stylisation of life' (Bourdieu, 1984), while only rapports between people and production (or consumption) are examined. Attention, in this volume, has been focused on the processes of aestheticisation and romanticisation and on the importance of identity creation and the role of age, race, gender, etc in constituting them. This text refutes Peet's assertion that "there is a yawning gap between the objectivity of economically derived logics and the extreme, even excessive, emotional subjectivity of agency/identity" (1997, p. 37).

This volume presents the broader and more imaginative understanding that highlights the diversity and complexity of relational networks of the new cultural economy and recognises that the economics of consumption (consumption being the engine of contemporary capitalism) are culturally constructed. The researchers have adopted the paradigm that landscapes and spaces translate and result from the intersections of multiple networks of relations; social, economic and cultural. The local is not an inevitable victim of global forces. The concepts of enworldment, unworldment, deworldment and transworldment evoke the interdependencies of the multiple facets and links of this new cultural economy of space. Crouch rightly wonders whether we are faced with a new understanding of processes, power and change by academia rather than a truly new form of spatial organisation. Academia does engage in discourses about the 'newness' of our time and concludes that modernity has created circumstances that cannot really be erased by new circumstances (Ben-Rafael & Sternberg, 2002, pp. 3-4). Old as well as new processes participate in this construction, fragmentation, reconstruction of our postmodern realities and landscapes and have been the subject of this volume, but at any time, certain facets of spatial organization will tend to be emphasised over others.

It is not possible to isolate single, one way, cause-effect links: Jauhainien points out that "… the body, the mind and the environment are particularly mixed and melted into a simultaneously present and absent here/t/here technologised landscape … We are connected to and part of these technologised landscapes through the ICT mediators" (p. 177). Myopic insistence on an abstract notion of capitalist globalisation or on specific isolated relations cannot solve problems that are in effect the consequences of these numerous links that operate at different scales between varied actors (singly or in groups), in multiple directions, affecting each other and the territories they cross in incalculable ways (Crouch, Sluyter).

Gibson Graham had posited that we must "confront and reshape the ways in which we live and enact the power of the global" (1997) but Allen also speaks of the 'roundaboutness' (2004, p. 23) of power. Foucault has effectively demonstrated that power engenders resistance too. The global, in other words, does not impose on the local or the individual that would have lost all agency, though such agency may be expressed in diverse forms. The global and the local (even down to the individual scale), and all levels in between, are mutually constituted and constitutive.

Bourdieu has decried globalisation as an assault of "wild capitalism" to retain Western economic domination of the world (d'Hauteserre, Geronimi), in the mould of colonial discourses about how "late precolonial peoples and their living descendants were incapable of having developed such a productive landscape. Only Pre-Columbian diffusions from the West could explain precolonial vestiges" (Sluyter, p. 103). Globalisation has effectively made the globe itself – or at least the term describes how – the relevant space of these interconnections (and) has led to the increased porousness between inside and outside but with varying intensities and strengths in different locations.

Even if bounded territoriality and identification with a specific location are problematic, local spaces have not been emptied by global flows and have not been transformed into more functionalist spaces (Crouch, Sluyter). Globalisation and the networks it engenders are anchored in specific places and participate in the construction of their landscapes. Economic behaviour is rooted in local settings generating tensions between local needs and extra local demands. Local cultural and cognitive forces create the social and regulatory arrangements within which local, as well as global, actors (embodied subjects, texts, machines, etc – see Callon 1991), individually or in groups, perform, creating materially heterogeneous economic geographies. Local attachment has always been defined through plural connections which are revealed in local landscapes as different networks coordinate the actors involved (Soini, Palang & Semm, Jauhainen). The authors demonstrate how, in turn, networks that define global reach have been constructed from spatially locatable values, symbols, products, practices and actors (Cosgrove, Olwig, Jauhainen). Global flows are not disconnected from specific locations and landscapes except perhaps in appearance. Authors of this volume agree that localities articulate in complex modes with the global as the "local level of life [remains] a crucial relational field" (Anderson 2001, p. 387).

Landscapes remain a major area of research as the consequence of the interconnections between networks and localities in the new context of increasing mobility. Different scales have been scrutinised as interactions between actors (in Callon's sense, 1991) and networks and their consequences vary according to the levels at which they occur. Local (Lange), regional (Cosgrove, Sluyter), and global (d'Hauteserre) levels in both urban (Geronimi, Olwig) and non-urban (Soini, Palang & Semm) settings have attracted scrutiny, though often entanglements between all of these levels have forced the researchers to include several levels in their narratives, particularly when they have discussed the lowest scale, that of the body or of the individual (Jauhainen, Crouch).

Authors in this volume offer an in-depth, often individualised engagement with different theoretical and methodological positions from diverse contexts, in order to advance the theorisation of the processes of enworldment, unworldment, deworldment and transworldment that effect and affect the contemporary transformations of landscapes as the interface between humans (culture, society) and the environment (socially constructed for/by economic and other necessities). Where the initial underlying implication of delving into these processes of the 'new cultural economy of space' may move, in many cases, in a scenario of 'placelessness', 'inauthenticity' or of dissolution of landscapes as known thus far, what emerges from all chapters is the conviction or the anticipation of new forms of landscape organization.

Whether explicitly described as such or not, these processes of 'reworldment' seem to describe in ever so fluid and evolving ways the products of the 'new cultural economy of space' in the contemporary landscape, carrying a wide range of implications, from positive to uneasy and elusive. New forms, functions, systems and networks referring to all aspects of life, forming new scalar spatialities, create new landscapes and places, 'new presences and absences' (Olwig). Olwig elaborates on the emergence of a complex multitude of political landscapes making up an anthropological space that, "like zero, may seem like nothing, but which actually constitute a nothing that is – an absent presence – and it has the potential to undermine the global hegemony" (p. 158). Some of the mechanisms at play may be new, as in the previous case, whereas others, such as effecting change on landscape through place-naming, may represent older, more ubiquitous practices, still relevant in place making (Soini, Palang & Semm).

There is increasing evidence that, despite the fact that in some cases local cultures are under pressures of homogenization, new forms and meanings of 'local' culture and identity are being produced (Jackson, 1999). Hints of reworldment are contained, for example, in the claiming by locals of their spaces, by asserting themselves through their choices and actions, as, for instance vis-à-vis tourism in non-heterotopic tourist destinations (d'Hauteserre).

Other authors also recognize additional characteristics of this 'new cultural economy of space' to those presented in the introduction of this volume, such as the significance of signs through which cultures are developed and communicated (Crouch), or tendencies of "segmentation, fusion and hybridity, ... leading to continuous cultural innovation" (Cosgrove, p. 71). Daniels and Cosgrove assert that the postmodern apprehension of the world emphasizes "the inherent instability of meaning, the ability to invert signs and symbols, to recycle them in a different context and thus transform their reference" (1988, p. 7-8). Elsewhere in the book, communications (both physical and social, internal and external) and the powerful presence of the individual in terms of entrepreneurial initiative in contexts of great diversity are cited as such qualities from which new forms of cultural economies of space germinate.

In fact, despite the general acknowledgment of tensions brought about by forces of contemporary change, the authors at times present a playful attitude towards processes of the 'new cultural economy of space' at hand, employing terms such as 'misworldment' and 'misdevelopment' (Sluyter), 'performing microglobalizations' (Lange) etc. More importantly, they see these processes and their products everywhere in the landscape, including landscapes of their personal geographies, suggesting an ongoing acknowledgement of, even reveling in, the richly imaginative ways in which people continue to live and 'world' their landscapes.

REFERENCES

Adkins, L & Lury, Celia. (1999). The labour of identity: performing identities, performing economies. *Economy and Society*, pp. 598-614.
Allen, J. (2004). The whereabouts of power: politics, government and space. *Geografiska Annaler*, 86B (1), pp. 19-32.

Anderson, K. (2001). Thinking postnationally. *Annals of the Association of American Geographers*, 90 (2), pp. 381-389.

Bataille, G. (1985). *Visions of excess: selected writings, 1927-39.* Manchester, Manchester University Press.

Ben-Rafael, E & Sternberg, Y. (2002). Analysing our time: a sociological problématique. In E. Ben-Rafael and Y. Sternberg (Eds.), *Identity, culture and globalisation*, pp. 3-19. Boston: Brill.

Bourdieu, P. (1984). *Distinction: a social critique of the judgement of taste.* London: Routledge.

Callon, M. (1991). Techno-economic networks and irreversibility. In J. Law, edr, *A sociology of monsters: essays on power, technology and domination*, pp. 132-161. London: Routledge.

Crang, P. (1997). Performing the tourist project. In C. Rojek and J. Urry (Eds) *Touring cultures: transformations of travel and theory.* London: Routledge.

Daniels, S and Cosgrove, D. (1988). Introduction: the iconography of landscape. In Denis Cogrove and Stephen Daniels (Eds.), *The iconography of landscape*, pp. 1-10. Cambridge: Cambridge University Press.

Gibson-Graham, J. K. (1997). Re-placing class in economic geographies: possibilities for a new class politics. In Roger Lee and Jane Willis (Eds), *Geographies of economies*, pp. 71-86. London: Arnold.

Jackson, Peter. (1999). Commodity culture: the traffic in things. *Transactions IBG*, NS 24, pp. 95-108.

Latour, B. (1991). Technology is society made durable. In J. Law, edr, *A Sociology of monsters: essays on power, technology and domination.* London: Routledge.

Massey, D. (1994*). Space, place and gender.* Cambridge: Polity Press.

Peet, R. (1997). The cultural production of economic forms. In Roger Lee and Jane Willis (Eds), *Geographies of economies*, pp. 37-46. London: Arnold.

Sheller, M. & Urry, J. (2000). The city and the car. *International Journal and Regional Research*, 24, pp. 737-757

Landscape Series

1. H. Palang and G. Fry (eds.): *Landscape Interfaces*. Cultural Heritage in Changing Landscapes. 2003   ISBN 1-4020-1437-6
2. I. Vanslembrouck and G. Van Huylenbroeck: *Landscape Amenities*. Economic Assessment of Agricultural Landscape. 2005   ISBN 1-4020-3134-3
3. F. Almo: *Principles and Methods in Landscape Ecology*. Towards a Science of Landscape. 2005   ISBN 1-4020-3327-3 Pb: ISBN 1-4020-3328-1
4. D.G. Green, N. Klomp, G. Rimmington and S. Sadedin: *Complexity in Landscape Ecology*. 2006   ISBN 1-4020-4285-X
5. T.S. Terkenli and A.-M. d'Hauteserre (eds.): *Landscapes of a New Cultural Economy of Space*. 2006   ISBN 1-4020-4095-4
6. Z. Naveh (ed.): *Transdisciplinary Challenges for Landscape Ecology and Restoration Ecology*. Anthology of Studies on Mediterranean and Global Issues. 2006   ISBN 1-4020-4420-8
7. Is not available yet.
8. F. Kienast, S. Ghosh, O. Wildi (eds.): *A Changing World*. Challenges for Landscape Research. 2006   ISBN 1-4020-4434-8